Studies in Space Policy

Volume 23

Series Editor

European Space Policy Institute, Vienna, Austria

The use of outer space is of growing strategic and technological relevance. The development of robotic exploration to distant planets and bodies across the solar system, as well as pioneering human space exploration in earth orbit and of the moon, paved the way for ambitious long-term space exploration. Today, space exploration goes far beyond a merely technological endeavour, as its further development will have a tremendous social, cultural and economic impact. Space activities are entering an era in which contributions of the humanities – history, philosophy, anthropology –, the arts, and the social sciences – political science, economics, law – will become crucial for the future of space exploration. Space policy thus will gain in visibility and relevance. The series Studies in Space Policy shall become the European reference compilation edited by the leading institute in the field, the European Space Policy Institute. It will contain both monographs and collections dealing with their subjects in a transdisciplinary way.

More information about this series at http://www.springer.com/series/8167

Annette Froehlich · Claudiu Mihai Tăiatu

Space in Support of Human Rights

ESPI
European Space Policy Institute

Springer

Annette Froehlich
European Space Policy Institute
Vienna, Austria

Claudiu Mihai Tăiatu
European Space Policy Institute
Vienna, Austria

ISSN 1868-5307 ISSN 1868-5315 (electronic)
Studies in Space Policy
ISBN 978-3-030-35425-1 ISBN 978-3-030-35426-8 (eBook)
https://doi.org/10.1007/978-3-030-35426-8

This Springer imprint is published by the registered company Springer Nature Switzerland AG
The registered company address is: Gewerbestrasse 11, 6330 Cham, Switzerland

Executive Summary

This book presents an analysis of human rights from the perspective of international space law and policy. It seeks to continue the analysis of how remote sensing activities and satellite data could be used in civil courts, the benefits of this technology, the challenges and future legal perspectives. Human rights refer to the basic rights and freedoms based on shared values such as dignity, fairness, equality, respect and independence. These values are defined and protected by law. The Universal Declaration of Human Rights contains in Article 1 the essence of the notion of human rights: "All human beings are born free and equal in dignity and rights. They are endowed with reason and conscience and should act towards one another in a spirit of brotherhood". Human rights should belong to every person in the world, from birth until death, applying regardless of where the individual comes from and believes in.

The satellite technology in Low Earth Orbit (LEO), in particular large constellations of satellites, is more affordable and capable of providing continued and global coverage with faster data transfer. Future space activities include the possibility of combining space-based data with ground-level activities in real time for quick analysis. Small satellites are in continuous development and LEO is likely to be a growing market for commercial space.

The Outer Space Treaty states in Article 1 that outer space is the province of mankind, to be used for the benefit of all humankind. While the scope of, or in fact the nature of an obligation that could be derived out of Article 1 has been the object of many, often controversial, discussions, there is a general understanding that space-based applications can aid in a number of ways, especially with regard to developing countries that often lack critical infrastructure. Looking at the world today, it seems that almost every day the news are filled with reports of disasters and images of refugees fleeing war or starvation. Developing countries are frequently affected by such setbacks, often lacking critical infrastructure to cope with such situations.

Satellite data can provide much needed relief and has already been used numerous times in legal proceedings to support judges in finding a fair judgment. Satellite data has been used by the International Court of Justice to determine the evolution of a border river in its judgment on the status of the Kasikili/Sedudu island,[1] thereby resolving a lasting territorial dispute. Furthermore, satellite data can be used to prove economic impacts, as shown in the "Certain Activities Carried out by Nicaragua in the Border Area" Case. The use of satellite imagery as evidence was also crucial in the "Oil Platforms" case, as it provided proof that Iran was using offshore oil platforms as missile sites to carry out attacks against neutral ships during the Iran-Iraq war. Recently, satellite data has been used to prove the violation of human rights—for example, to demonstrate the demolition of housing in Zimbabwe.

Finally, satellite data may also provide evidence of war crimes. Based on pre-attack and post-attack satellite images, the damage to the MGS Haskanita military base in northern Darfur and thereby the unlawful attack against peacekeepers of the African Union could be verified. Small satellites can prove to be a viable solution for developing countries, as their low mass and size (below 500 kg) helps to reduce the economic cost for launch vehicles and the costs associated with construction—thereby turning them into an affordable solution to create much needed telecommunication infrastructure and obtain critical satellite data for developing countries to improve security, development, stability, justice and peace.

It is important to disseminate information about space surveillance, in particular human-based activity monitoring, because it has the potential to become an essential tool for judges in reaching a fair judgment, to prove the violation of human rights and provide evidence of war crimes.

The rationale for this project is derived from the worrying scale and intensity of conflicts, humanitarian crises and human rights violations around the world. This extends to a wide range of global hotspots including Venezuela, Yemen, Syria, Myanmar, Sudan, Eritrea and numerous other locations. These developments also have relevance for Europe, given large-scale migration. In response to these developments, it becomes imperative to analyse ways in which space can contribute to upholding human rights and monitoring violations. Various international tribunals, such as the International Court of Justice (ICJ) and the International Criminal Court (ICC), are increasingly relying on satellite data and especially images when considering cases related to human rights violations.

This use of space-related technological developments is a trend that promises to continue as the range and accuracy of space-derived data improves. Satellite data have important legal implications because they enable the monitoring of international obligations and are a powerful tool for dispute resolution. Therefore, this publication will take into consideration the use of satellite images for cases related to human rights violations, since the multitude of various humanitarian crises worldwide demonstrate that it is of the utmost interest to analyse how space law, policies and space-related applications could further support the implementation and monitoring of the observance of human rights, thus contributing to greater security and sustainable development. A range of relevant areas, such as migration,

[1] ICJ -International Court of Justice, Judgment of 13 December 1999, Paragraph 25, pp. 21 https://www.icj-cij.org/files/case-related/98/098-19991213-JUD-01-00-EN.pdf (accessed 06.09.2019).

refugees (including settlements and whether they are supplied adequately with basic necessities), water distribution and quality, housing and settlement monitoring, form crucial aspects to be considered in this study. Critically, the use of satellite data for legal purposes is not without problems or concerns, and these will also be considered to aid future strengthening of this developing field.

Contents

About the Authors

Dr. Annette Froehlich is a scientific expert seconded from the German Aerospace Center (DLR) to the European Space Policy Institute (Vienna), and Honorary Adjunct Senior Lecturer at the University of Cape Town (SA) at SpaceLab. She graduated in European and International Law at the University of Strasbourg (France), followed by business-oriented postgraduate studies and her Ph.D. at the University of Vienna (Austria). Responsible for DLR and German representation to the United Nations and International Organizations, Dr. Froehlich was also a member/alternate head of delegation of the German delegation to UNCOPUOS. Moreover, Dr. Annette Froehlich is author of a multitude of specialist publications and serves as a lecturer at various universities worldwide in space policy, law and society aspects. Her main areas of scientific interest are European Space Policy, International and Regional Space Law, Emerging Space Countries, Space Security and Space & Culture. She has also launched as editor the new scientific series "Southern Space Studies" (Springer publishing house) dedicated to Latin America and Africa. Annette.Froehlich@espi.or.at; Annette.Froehlich@dlr.de.

Claudiu Mihai Tăiatu is a Romanian lawyer. He graduated in 2017 from the Adv. LL.M. of Air and Space Law of the International Institute of Air and Space Law (IIASL), Leiden University, the Netherlands and in 2018 from the International Space University (ISU), Space Studies Program (SSP18). In 2017 he was awarded the International Institute of Space Law (IISL) Prof. Dr. I.H.Ph. Diederiks-Verschoor Award for his research on Space Traffic Management. In 2018 he was awarded in the Worldwide Space Law Essay Competition "Legal Aspects Relating to Satellite Constellations" organized by the ECSL, ESPI and DLR. He has successfully completed several internships at the Regulatory Affairs Department of OneWeb, the ITU Radiocommunication Bureau, ESPI and UNIDROIT. He is part of the SGAC Space Law and Policy Newsletter Team.

The authors specially acknowledge the assistance of Rutwik B. Navalgund, who recently graduated from the Master of Air and Space Law programme at the International Institute of Air and Space Law (IIASL), Leiden University, for his contribution to this research with an analysis of monitoring compliance in connection with human rights, the use of remote sensing in various Multilateral Environmental Agreements (MEAs), and a presentation of the important agreements that utilize satellite data.

Acronyms and Abbreviations

AAAS	American Association for the Advancement of Science
AEM	Mexican Space Agency (*Agencia Espacial Mexicana*)
AI	Amnesty International
AI	Artificial Intelligence
AIS	Automated Identification System
ANN	Artificial Neural Networks
ANPRM	Advance Notice of Proposed Rulemaking
ASA	African Space Agency
AU	African Union
CCAMLR	Convention for the Conservation of Antarctic Marine Living Resources
CEP	Committee on Environmental Protection
CNES	National Centre for Space Studies *(Centre National d'Études Spatiales)*
CNN	American news—Cable News Network
COIP	The Integral Organic Criminal Code
COPUOS	Committee on the Peaceful Uses of Outer Space
CSA	Canadian Space Agency
CVPD	Chula Vista Police Department
DEM	Digital Elevation Model
DIAS	Data and Information Access Services
DMC	Disaster Monitoring Constellation
DRC	Democratic Republic of the Congo
DTM	Digital Terrain Model
EAfROAD	East Africa Regional Office of Astronomy for Development
EARSC	European Association of Remote Sensing Companies
ECHR	European Convention on Human Rights
ECJ	Court of Justice of the European Union
ECV	Essential Climate Variables
EMSA	European Maritime Safety Agency

EO	Earth Observation
EOMAP	Earth Observation and Environmental Services
EOSDIS	NASA's Earth Observing System Data and Information System
ESA	European Space Agency
ESPI	European Space Policy Institute
ESSTI	Ethiopian Space Science and Technology Institute
EU LIBE	European Parliament Committee on Civil Liberties, Justice and Home Affairs
EU	European Union
EUMETSAT	European Organization for the Exploitation of Meteorological Satellites
FAA	Federal Aviation Administration
FAO	United Nations Food and Agriculture Organization
FAW	Peninsula located in southern Iraq, where the Euphrates and Tigris rivers meet
GCOS	Global Climate Observing System
GCP	Ground Control Points
GDPR	The European Union General Data Protection Regulation
GFSC	Global Framework for Climate Services
GFW	Global Fishing Watch
GHGs	Greenhouse Gases
GIS	Geographic Information Systems
GNSS	Global Navigation Satellite System
GO	Governmental Organisation
GPS	Global Positioning System
GRID19	Global Report on Internal Displacement Issued by Internal Displacement Monitoring Centres of the United Nations
GSICS	Global Space-based Inter-Calibration Systems
GSO	Geostationary Orbit
HR	Human Rights
IAC19	70th International Astronautical Congress, 21–25 October 2019
IAU	International Astronomical Union
IBRU	International Boundaries Research Unit
ICC	United Nations International Criminal Court
ICCPR	International Covenant on Civil and Political Rights
ICJ	United Nations International Court of Justice
ICMP	International Commission on Missing Persons
ICT	Information and Communication Technologies
ICTY	United Nations International Criminal Tribunal for the former Yugoslavia
IDMC's	Internal Displacement Monitoring Centres of the United Nations
IDPs	Internally Displaced Persons
IED	Improvised Explosive Device
IGO	Inter-Governmental Organization
IMAZON	Institute of Man and Environment of the Amazon

INPE	Brazilian National Institute for Space Research (*Instituto Nacional de Pesquisas Espaciais*)
IO	International Organisation
IPAM	The Institute for Amazon Environmental Research of Brazil
IPP	Integration Pilot Program
ISI	ImageSat International
ISMA	International Satellite Monitoring Agency
ISRO	Indian Space Research Organization
ISS	International Space Station
ITLOS	International Tribunal for the Law of the Sea
ITU	International Telecommunication Union
IUU Fishing	Illegal, Unreported and Unregulated (IUU) fishing
IWGIA	International Work Group for Indigenous Affairs
JAXA	Japanese Space Agency
LANDSAT	A US scientific satellite that studies and photographs the Earth's surface by using remote sensing techniques
LEO	Low Earth Orbit
MEAs	Multilateral Environmental Agreements
MEXSAT	Mexican Satellite System
MGS	Haskanita Military Group Site
MODIS	Moderate Resolution Imaging Spectroradiometer
MV NIKA	Panama-flagged cargo vessel—750 tonnes with a long track record as an outlaw vessel
NARSS	National Authority for Remote Sensing and Space Sciences
NASA	United States National Aeronautics and Space Administration
NGOs	Non-Governmental Organizations
NOAA	United States National Oceanic and Atmospheric Administration
NPRM	Notice of Proposed Rulemaking
NWP	Numerical Weather Prediction
OHCHR	Office of the High Commissioner for Human Rights
OSCE	Organization for Security and Co-operation in Europe
OST	Outer Space Treaty
RtoP	Responsibility to Protect
SA Catapult	Satellite Applications Catapult
SAF	Syrian Air Forces
SAR	Synthetic Aperture Radar
SDGs	United Nations Sustainable Development Goals
SKA	Square Kilometre Array
SNHR	Syrian Network for Human Rights
SPOT	A commercial high-resolution optical imaging Earth observation satellite system operating from space
SRS	Satellite Remote Sensing
STRM	Shuttle Radar Topographic Mission
TEU	Treaty of the European Union
UAE	United Arab Emirates

UAS	Unmanned Aerial Systems
UAV's	Unmanned Aerial Vehicles
UDHR	Universal Declaration of Human Rights
UK	The United Kingdom of Great Britain and Northern Ireland
UN Charter	Charter of the United Nations
UN MINUSCA	United Nations Multidimensional Integrated Stabilization Mission in the Central African Republic
UN	The United Nations
UNCLOS	United Nations Convention on the Law of the Sea
UNESCO	United Nations Educational, Scientific and Cultural Organization
UNGA	United Nations General Assembly
UNHCR	United Nations High Commissioner for Refugees
UNICEF	United Nations International Children's Emergency Fund
UNITAR	United Nations Institute for Training and Research
UNODC	United Nations Office on Drugs and Crime
UNOOSA	United Nations Office for Space Affairs
UNOSAT	UNITAR's Operational Satellite Applications Programme
UNPoA	United Nations Programme of Action to Prevent, Combat and Eradicate the Illicit Trade in Small Arms and Light Weapons in All Its Aspects
USA	United States of America
USDOT	United States Department of Transportation
VDPA	Vienna Declaration and Programme of Action
VHR	Very High Resolution
VIIRS	Visible Infrared Imaging Radiometer Suite
VMS	Vessel Monitoring System Washington, D.C.
WFP	World Food Programme
WSSD	World Summit on Sustainable Development, Johannesburg Summit, Johannesburg, South Africa, 26 August–04 September 2002

Chapter 1
Human Rights and Space Law

Abstract This chapter starts by discussing human rights in general. Information will include the Universal Declaration of Human Rights and the Charter of the United Nations but also an analysis of the contributions of technology to supporting human rights. Further, the relationship of Space Law to Human Rights is discussed as well as the UN Principles on Remote Sensing and the importance of distinguishing between raw data and processed data is presented. An overview of the challenges for considering satellite data as evidence in courts, including questions about the authenticity of processed data is provided. The lack of consistency and standardization in processing the data is discussed and the requirement for a competent authority, international organization or national agency, with certification power is underlined. Furthermore, the capacity of satellite images to improve probatory procedures in the judicial system and prove violations of human rights and international humanitarian law is analysed. Finally, this chapter will discuss the challenges that pertain to the interpretation of satellite images necessary for the prosecution of criminals.

1.1 Introduction to Human Rights

Human rights refer to the basic rights and freedoms to which all humans are entitled. Their realization depends on "the contributions that each and every one of us is willing to make, individually and collectively, now and in the future".[1] Their recognition is "the foundation of freedom, justice and peace in the world".[2] Published in 1762, *The Social Contract* of Jean-Jacques Rousseau inspired political reforms in Europe and asserted the people are sovereign in choosing the rights under which they live and argued "that force does not create right, and that we are obliged to obey only legitimate powers".[3] Thus, developed the idea of a general will (volonté générale)

[1] United Nations Human Rights Office of the High Commissioner, *Human Rights day 2004*, https://www.ohchr.org/EN/NewsEvents/HRDay2004/Pages/objectives.aspx, (accessed 11.05.2019).

[2] UN General Assembly, Universal Declaration of Human Rights, 10 December 1948, 217 A (III), Preamble, https://www.un.org/en/universal-declaration-human-rights/, (accessed 11.05.2019).

[3] Rousseau, J.J., The Social Contract, Book one, Chapter three, CreateSpace Independent Publishing Platform, March 7, 2014. See also: https://www.coursehero.com/lit/The-Social-Contract/book-1-chapters-1-6-summary/, (accessed 23.09.2019).

A. Froehlich and C. M. Tăiatu, *Space in Support of Human Rights*, Studies in Space Policy 23, https://doi.org/10.1007/978-3-030-35426-8_1

and the private interest (intérêt particulier) of an individual which in this case "may dictate to him very differently from the common interest".[4] For Rousseau, "Man loses by the social contract his *natural* liberty (..) in return he acquires *civil* liberty, and proprietorship of all he possesses". The UN High Commissioner for Human Rights emphasized that human rights moved on from the idea of individual privileges in a national revolutionary proclamation, such was the case for example in the French Declaration of 1789 or the political settlements contained in the Magna Charta of 1215, to the highest international principles relevant to support better government and fight injustice, discrimination and inhumanity.[5]

The protection of human rights in international law dates from the end of the World War II and the creation of the UN.[6] Under the League of Nations established at the end of the World War I, the international community attempted to develop a system of protection for human rights and international monitoring systems.[7] The establishment of the UN in 1945 represents the beginning of an unprecedented international protection of human rights and thus only after the World War II was such an international system established.

The Charter of the United Nations (UN Charter) of 1945 is the foundational treaty of the UN. The UN Charter articulates the commitment to upholding the human rights of citizens and outlines a broad set of principles relating to achieving "higher standards of living", addressing "economic, social, health, and related problems" and "universal respect for, and observance of, human rights and fundamental freedoms for all without distinction as to race, sex, language, or religion". The UN Charter is a constituent treaty, and all members are bound by its articles. According to the provisions of Article 103 of the UN Charter, obligations to the UN prevail over all other treaty obligations. The UN Charter makes reference to human rights in its preamble "to reaffirm faith in fundamental human rights, in the dignity and worth of the human person, in the equal rights of men and women (…)" and also in Article 1 (3) when proclaiming "respect for human rights and for fundamental freedoms" as the purpose of the UN. In Article 55 (c) the UN pledges as an organization to promote "universal respect for, and observance of, human rights and fundamental freedoms for all without distinction as to race, sex, language or religion", and in Article 76 (c) it aims "to encourage respect for human rights and for fundamental freedoms for all (…)". The UN General Assembly adopted the Universal Declaration of Human

---

[4] Clapham, A., Human Rights, A Very Short Introduction, Oxford University Press, 2007, pp. 17–19, ISBN 978-0-19-920552-3.

[5] Ibid. *supra* note Clapham, A., Human Rights, A Very Short Introduction, Oxford University Press, 2007, pp. 17–19, ISBN 978-0-19-920552-3.

[6] McClean, E., The dilemma of intervention: Human rights and the UN Security Council, in: Marco Odello and Sofia Cavandoli (Eds.), Emerging Areas of Human Rights in the 21st Century. The Role of the Universal Declaration of Human Rights, New York, Routledge, 2011, pp. 25.

[7] United Nations, The United Nations Human Rights Treaty System: An introduction to the core human rights treaties and the treaty bodies. Fact sheet No.30/ 2005 United Nations Office in Geneva. https://www.ohchr.org/documents/publications/factsheet30rev1.pdf, (accessed 11.05.2019). See also O.M. Enerstvedt, Aviation Security, Privacy, Data Protection and Other Human Rights, pp. 67.

Rights (UDHR) on 10 December 1948. Together with the UN Charter, these are the basic texts for the protection of human rights.[8]

The UNGA decided by Resolution 2081 to convene an International Conference on Human Rights to promote the principles contained in the Universal Declaration of Human Rights, to develop and guarantee political, civil, economic, social and cultural rights and to end all discrimination and denial of human rights and fundamental freedoms on grounds of race, colour, sex, language or religion, "and in particular to emphasize the urgent need to eliminate discrimination and other violations of human dignity, with special attention to the abolition of racial discrimination and in particular the policy of *apartheid*".[9] The 1968 International Conference on Human Rights convened by the UN in Tehran to celebrate the International Year for Human Rights was the first global meeting on human rights. The conference was held from 22 April to 13 May 1968 to review the progress made in the 20 years since the adoption of the Universal Declaration of Human Rights and formulate an agenda for the future.[10] The conference was attended by representatives of 84 states and by representatives or observes from a number on UN bodies and specialized agencies, inter-governmental (IOs) and non-governmental organizations (NGOs) and concluded with the adoption of the "Proclamation of Tehran" contained in the Final Act of the International Conference on Human Rights (A/CONF.32/41).[11]

Ever since the first UN Conference on Human Rights held in 1968 in Tehran, there has been an intrinsic relationship between development and human rights.[12] The content of human rights and the inherent natural rights is now understood by reference to the international treaties. At international level human rights operate from a higher plane in the hierarchy than general law, for example certain international law prohibitions (known as "peremptory" or *jus cogens* norms) are of superior strength.[13] In general, the national constitutions and domestic laws include provisions that stipulate that in case of conflict between the provisions of national law and the international law, the provisions of international treaties shall prevail. Another aspect for recognizing human rights is the incorporation into domestic laws of the international instruments. By doing so, this significantly strengthens the scope and effectiveness

---

[8]Odello, M., Cavandoli, S. (Eds.), Emerging Areas of Human Rights in the 21st Century. The Role of the Universal Declaration of Human Rights, New York, Routledge, pp. 1, 2011.

[9]UN GA Resolution 2081 (xx), International Year for Human Rights, UN doc. A/RES/2081(XX), 20 December 1965.

[10]International Conference on Human Rights, Teheran, 22 April–13 May 1968, See Thomson, A.S., Teheran 1968 and the Origins of the Human Rights Council? https://acuns.org/wp-content/uploads/2012/06/Teheran-Origin-Human-Rights-Council-Andrew-Thompson-AM-2011.pdf, (accessed 11.05.2019). See also: https://www.un.org/en/development/devagenda/humanrights.shtml.

[11]United Nations Audiovisual Library of International Law, http://legal.un.org/avl/pdf/ha/fatchr/fatchr_ph_e.pdf, (accessed 11.05.2019).

[12]Verna, S.K., Development, Human Rights and Globalisation of Economy, in: Journal of the Indian Law Institute, Vol. 40, No. 1/4, Human Rights Special Issues, pp. 217–229, 1998, https://www.jstor.org/stable/43953318?seq=1#page_scan_tab_contents, (accessed 11.05.2019).

[13]Ibid. *supra* note Andrew Clapham, pp. 18.

of measures and becomes binding. It enables courts to adjudicate violations of the rights by direct reference to international treaties.

It is noteworthy to underline that the advocacy of human rights ideas by NGOs is not unusual as it is considered an important element for the creation and application of human rights protection. The number of advocacy groups for human rights grew significantly in the twentieth century even if the core group regarding international law of human rights and humanitarian affairs remained relatively small (171 states and more than 800 NGOs attended the World Conference on Human Rights, 14–15 June 1993, Vienna Austria).[14]

Amnesty International (AI), Human Rights Watch, and Physicians for Human Rights, are highly active NGOs in human rights matters and generate some influence and pressure on states to do the right things.[15] In general the international law of human rights is mainly a state agenda problem but the role of private non-profit organizations like human rights advocacy groups is also important as they help shape the rights discourse and action. Their lobbying efforts results influence policy in the short term through dissemination and in the long term as "those educated today may be the policy makers of tomorrow".[16] AI is a globally known organization on the advocacy of traditional human rights. Since its founding in 1961 by Beneson, it has developed a general reputation for accurate reporting primarily about "prisoners of conscience" (imprisoned for their social and political views, physically restrained from expressing any opinion) but also about torture and the death penalty, working for "fair and prompt trials for all political prisoners".[17] AI cooperates in many of its projects with the ICJ and other NGOs such as Human Rights Watch. It has the ability to influence human rights norms based on its attributes of legitimacy such as: loyalty to and promotion of the principles of human rights, with a focus on the "prisoners of conscience"; independence and political impartiality, playing a role of disinterested and autonomous third party in the international system; but also using information in the service of the general need for norms, its interpretative capacity represents the ability to determine new concepts about human rights based on collected facts.[18] The actions of AI emphasize that the rights of individuals need protection under any type of government, underlining its apolitical mandate.

The interpretive capacity of AI is of fundamental importance as it refers to the interpretation of facts based on information about behaviour. Interpretation is much more than just the collection of facts as it offers a critical assessment on human rights violations based on monitoring of activities at a given period and place. As was already pointed out, because human rights violations are so often hidden, information

[14]United Nations Human Rights Office of the High Commissioner, https://www.ohchr.org/en/aboutus/pages/viennawc.aspx, (accessed 11.05.2019).

[15]Forsythe, D.P., Human Rights in international relations, pp. 153, Third Edition, Cambridge University Press, 2006.

[16]Ibid. *supra* note Forsythe, D.P., Human Rights in international relations, pp. 193, Third Edition, Cambridge University Press, 2006.

[17]Power, J., Amnesty International. The Human Rights story, pp. 21 Pergamon Press, 1981.

[18]Clark, A.M., Diplomacy of Conscience Amnesty International and Changing Human Rights Norms, pp. 11, Princeton University Press, 2001.

is not always available on demand, is not easily acquired either within borders or across borders, and is definitely not cheap to collect.[19] As a consultative body to the UN, its role and the expertise of the NGOs in human rights monitoring, which is often more specialized than the expertise of states or IOs, has been acknowledged. UDHR is the only document that is explicitly mentioned in the AI mission statement which states that the organization "works to promote all the human rights enshrined in the Universal Declaration of Human Rights and other international standards through human rights education programs and campaigning for ratification of human rights treaties".

With emerging technologies, the future of human rights is not easy to predict. Nevertheless, human rights in international relations have to be protected and the positive contributions of technology should support human rights and social benefits. Any use of space technology, which could potentially interfere with the classical meaning of privacy, should be justified, such as being for the purpose of proving criminal offences or other illegal actions. The meaning of absolute rights and if or which rights could be limited in some cases and in which proportion in order for such an intrusion to benefit the society as a whole, must be analysed. Challenges to human rights have to be analysed in relation to the utilization of digital technology and the issues related to the current expansion of the human activities in space. Even if the digital age could not have been foreseen in 1948 by the drafters of the UDHR, its principles remain the fundamentals of human rights and represent a clear framework to secure the dignity and rights of all people.[20]

The Vienna Declaration and Programme of Action (VDPA) is a human rights declaration adopted by consensus at the World Conference of Human Rights on 25 June 1993 in Vienna, Austria. It states that ".... the promotion and protection of human rights is a matter of priority for the international community, and that the Conference affords a unique opportunity to carry out a comprehensive analysis of the international human rights system and of the machinery for the protection of human rights, in order to enhance and thus promote a fuller observance of those rights, in a just and balanced manner".[21] The VDPA reaffirms "the commitment to the purposes and principles" contained in the Charter of the UN and the UDHR. The VDPA seeks to reaffirm human rights as universal and relevant standards in its preamble: "Emphasizing that the Universal Declaration of Human Rights, which constitutes a common standard of achievement for all peoples and all nations, is the source of inspiration and has been the basis for the United Nations in making advances in standard setting as contained in the existing international human rights

---

[19] Ibid. *supra* note Clark, A.M., Diplomacy of Conscience Amnesty International and Changing Human Rights Norms, pp. 16.

[20] UN Human Rights Office of the High Commissioner, The Universal Declaration of Human Rights at 70: Still Working to ensure freedom, Equality and Dignity for all, https://www.standup4humanrights.org/layout/files/30on30/UDHR70-30on30-article12-eng.pdf, (accessed 11.05.2019).

[21] Vienna Declaration and Programme of Action, Adopted by the World Conference on Human Rights in Vienna on 25 June 1993, https://www.ohchr.org/Documents/ProfessionalInterest/vienna.pdf, (accessed 11.05.2019).

instruments, in particular the International Covenant on Civil and Political Rights and the International Covenant on Economic, Social and Cultural Rights." The VDPA considered "the adaptation and strengthening of the United Nations machinery for human rights" and recommended in Paragraph 18 of its Preamble the establishment of the position of the UN High Commissioner for Human Rights.

The Office of the High Commissioner for Human Rights (OHCHR) is the leading UN entity on human rights. The mandate of OHCHR derives from Articles 1, 13 and 55 of the UN Charter, the Vienna Declaration and Programme of Action, and UNGA Resolution 48/141 of 20 December 1993 (A/RES/48/141),[22] by which the General Assembly established the post of the UN High Commissioner for Human Rights. In connection with the program for reform of the UN (A/51/950, Paragraph 79), the OHCHR and the Centre for Human Rights were consolidated into a single OHCHR on 15 September 1997. The OHCHR is part of the UN Secretariat, with a unique mandate to promote and protect the human rights that are guaranteed under international law and stipulated in the UDHR.[23] The headquarters was established by the UN General Assembly on 20 December 1993 in the wake of the 1993 World Conference on Human Rights. The office is headed by the High Commissioner for Human Rights, who coordinates human rights activities throughout the UN System and acts as the secretariat of the Human Rights Council in Geneva, Switzerland.

[22] A/RES/48/141, 85th plenary meeting 20 December 1993, High Commissioner for the promotion and protection of all human rights https://www.refworld.org/docid/3b00f29b14.html, (accessed 11.05.2019).

[23] Mandate of OHCHR:

(a) *To promote and protect the effective enjoyment by all of all civil, cultural, economic, political and social rights;*

(b) *To carry out the tasks assigned to him/her by the competent bodies of the United Nations system in the field of human rights and to make recommendations to them with a view to improving the promotion and protection of all human rights;*

(c) *To promote and protect the realization of the right to development and to enhance support from relevant bodies of the United Nations system for this purpose;*

(d) *To provide, through the Centre for Human Rights of the Secretariat and other appropriate institutions, advisory services and technical and financial assistance, at the request of the State concerned and, where appropriate, the regional human rights organizations, with a view to supporting actions and programmes in the field of human rights;*

(e) *To coordinate relevant United Nations education and public information programmes in the field of human rights;*

(f) *To play an active role in removing the current obstacles and in meeting the challenges to the full realization of all human rights and in preventing the continuation of human rights violations throughout the world, as reflected in the Vienna Declaration and Programme of Action;*

(g) *To engage in a dialogue with all Governments in the implementation of his/her mandate with a view to securing respect for all human rights;*

(h) *To enhance international cooperation for the promotion and protection of all human rights;*

(i) *To coordinate the human rights promotion and protection activities throughout the United Nations system;*

(j) *To rationalize, adapt, strengthen and streamline the United Nations machinery in the field of human rights with a view to improving its efficiency and effectiveness;*

(k) *To carry out overall supervision of the Centre for Human Rights.*

Human rights is one of the three pillars of the UN Charter along with peace and development.[24] Even so, it is with respect to human rights that the UN's performance has experienced some of its greatest shortcomings, meaning the human rights pillar receives only a fraction of the resources enjoyed by the other two, a mere three percent of the general budget.[25]

The UN Sustainable Development Goals (SDGs) are particularly relevant for human rights. In terms of peace, justice and human rights, SDG16 (Peace, Justice and Strong Institutions) addresses the promotion of peaceful and inclusive societies for sustainable development, providing access to justice for all and building effective, accountable and inclusive institutions at all levels. More specifically, it ensures public access to information and protects fundamental freedoms in accordance with national legislation and international agreements by strengthening relevant national institutions, through international cooperation, capacity building (particularly in developing countries), preventing violence, and combating terrorism and crime.[26] SDG 17 (Partnerships for the Goals) promotes the strengthening of the various means of implementation and revitalization of the global partnership for sustainable development.[27] In this context, data from Earth Observation (EO) satellites can help in gathering evidence of international crimes, especially areas where access may otherwise be difficult, if not impossible.[28]

1.1.1 *Universal Declaration of Human Rights and Historical Development Before*

By its resolution 217 A (III) of 10 December 1948, the UN General Assembly, meeting in Paris, adopted the UDHR.[29] The General Assembly proclaimed the UDHR of 1948 as "a common standard of achievement for all peoples and all nations". The UDHR can be regarded as an instrument of the highest importance for the protection

---

[24] Dictum during the World Summit of 2005: "There can be no peace without the development, no development without peace, and neither without respect for human rights".

[25] Gilmour, A., *The Future of Human Rights: A view from the United Nations*, 2014 in Ethics International affairs, 28, no. 2, pp. 239–250, https://www.un.org/sg/sites/www.un.org.sg/files/atoms/files/Gilmour-FutureOfHumanRights.pdf, (accessed 23.09.2019).

[26] Martin, A.S., Satellite Data as Evidences Before the Mechanism of International Courts, in: Annette Froehlich (Eds.) Post 2030-Agenda and the Role of Space: The UN 2030 Goals and Their Further Evolution Beyond 2030 for Sustainable Development, Springer International Publishing AG, part of Springer Nature 2018.

[27] Ibid. *supra* note Martin, A.S., "Satellite Data as Evidences Before the Mechanism of International Courts".

[28] Ibid. *supra* note Martin, A.S., "Satellite Data as Evidences Before the Mechanism of International Courts".

[29] Ibid. *supra* note Odello, M, Cavandoli, S., (Eds.), Emerging Areas of Human Rights in the 21st Century. The Role of the Universal Declaration of Human Rights, Routledge, pp. 1, 2011.

and promotion of the rights of individuals and the furtherance of peace and stability.[30] This is the first intergovernmental statement to approve a set of basic principles on universal human rights "for all people and all nations".[31] The UDHR was solemnly described as the "moral anchor" of the worldwide movement in combating human rights violations,[32] but also as an "historic act, destined to consolidate world peace through the United Nations".[33] It has affirmed the principles that human beings shall enjoy fundamental rights and freedoms without discrimination and was expressed as a "common standard of achievement for all peoples and all nations".[34]

The UDHR is not a legally binding instrument and contains a wide span of rights covering all aspects of life divided into specific groups of rights such as civil, cultural, economic, political and social.[35] Since 1948, the proclamation of rights has experienced modifications by dozens of treaties and intergovernmental declarations that supplemented its provisions. The expanding and ever-evolving nature of human rights has raised the need to develop a complex system of human rights, enforced both at regional and global level. Member states and not only they, have an obligation to promote human rights and must refrain from interfering with or curtailing the enjoyment of human rights. The UDHR encourages states to adopting the necessary national instruments in supporting human dignity, promoting standards of life, respecting standards of life and ensuring external recognition. This is the reason why such protection includes the obligation to ensure that non-state parties do not infringe on human rights.[36]

Current topical discussion on human rights in relation to new space activities is related to satellite-based data such as images and video from remote sensing. These relative recent developments could not have been anticipated when the UDHR was signed. The preamble of the UDHR provides a strong statement emphasizing the core notions of freedom, dignity and equality as inherent features of humankind. The UDHR does not specifically refer to space activities. The significance of the UDHR is its inclusive language setting broad guidelines and general principles.

---

[30]UNGA Resolution 2081 (xx), International Year for Human Rights, UN doc. A/RES/2081(XX), 20 December 1965.

[31]Ibid. *supra* note Forsythe, D.P., Human Rights in international relations, pp. 152, Third Edition, Cambridge University Press, 2006.

[32]Morsink, J., The Universal Declaration of Human Rights. Origins, Drafting, and Intent, pp. xii Introduction, University of Pennsylvania Press, 1999.

[33]International Bill of Human Rights, Universal Declaration of Human Rights, 8 December 1948, https://www.un.org/en/ga/search/view_doc.asp?symbol=A/RES/217(III), (accessed 23.09.2019).

[34]UNGA, Convention Relating to the Status of Refugees, 28 July 1951, United Nations, Treaty Series, vol. 189, p. 137. Preamble, https://www.unhcr.org/3b66c2aa10, (accessed 23.09.2019).

[35]United Nations. The United Nations Human Rights Treaty System: An introduction to the core human rights treaties and the treaty bodies. Fact sheet No. 30. United Nations Office in Geneva, 2005, pp. 5.

[36]Enerstvedt, O.M., Aviation security, Privacy, Data Protection, and Other Human Rights: Technologies and Legal Principles, Issues in Privacy and Data Protection Series Volume 37, Springer International Publishing, 2017.

Another example where the UDHR is not specific is virtual space, the Internet and the ability to perform critical communications through e-mail. National laws have instead an important function to deal with cyber space as an arena where freedom of expression was shaped differently than the conventional boundaries, with a different understanding of restrictions, crime and offences. Digital privacy is not an abstract concept and different laws address it. A global example of digital interference and privacy of individuals is Facebook. Physical privacy is however a different concept that could be also violated by digital means such as remote sensing. This is the reason why the right of privacy in the digital age needs to include the particularity that individuals are not always aware of what they are surrounded by and should have the means to legally react to violation of privacy.

The right to non-discrimination and the right to equality are fundamental norms of international human rights law. They both refer to equality rights with the distinction that the right to non-discrimination is narrower in content and subsumed in the right to equality.[37] It was analysed in the doctrine that equality enjoys a triple status in the UDHR as (i) a general principle; (ii) an autonomous right; (iii) an accessory right. The preamble of the UDHR in the first recital refers to the "equal and inalienable rights of all members of the human family" but also the fifth recital strongly reaffirms the "equal rights of men and women". The general principle of the human rights framework is further stated in Article 1 UDHR when referring to equality as an inherent right of all human beings. The provisions of Article 2 attach the right of non-discrimination to all the other human rights set forth in the UDHR and, together with the provisions of Article 7 second sentence, recognizes accessory rights and affirms non-discrimination as an accessory right to equality.[38] The provisions describing an autonomous right of "equality before the law" and "equal protection of the law" are enshrined in the first sentence of Article 7. This freestanding right is valuable in the context of the lack of relevant rights in some jurisdictions.[39]

The term "privacy" used in Article 12 UDHR and in Article 17 of the International Covenant on Civil and Political Rights (ICCPR)—or "private life" used in Article 8 of the European Convention on Human Rights (ECHR) does not explicitly mention the aspect of privacy, but merely represents a general term, an "umbrella" notion that is so broadly stated that it offers protection to aspects of privacy not mentioned or not even foreseen during the codification process.[40] "Privacy is not an absolute right" and

---

[37] Appendix E to the Report of the Global Citizenship Commission "Article 7: The Equality and Non-Discrimination Provision".

[38] Ibid. Appendix E to the Report of the Global Citizenship Commission "Article 7: The Equality and Non-Discrimination Provision".

[39] *Article 7 All are equal before the law and are entitled without any discrimination to equal protection of the law. All are entitled to equal protection against any discrimination in violation of this Declaration and against any incitement to such discrimination.*

[40] Diggelmann, O., Cleis, M.N., How the Right to Privacy Became a Human Right, Human Rights Law Review, Volume 14, Issue 3, September 2014, pp. 441–458, https://doi.org/10.1093/hrlr/ngu014, (accessed 3.09.2019).

itself may be limited in some cases.[41] Intrusions on privacy must be legal, meaning in proportion to the benefit of the society as a whole. Otherwise, the provisions of Article 12 UDHR are very clear in limiting arbitrary interference to privacy, family, home or correspondence.[42] The right to privacy is one of the core elements of the UDHR and is essential for the protection of human dignity and free development of the human personality. In general, privacy can be understood as the right to protect individuals by clearly defining who will be granted legal access to their personal activity, communications and information. The core of the right to privacy contains at least two competing ideas: one is about privacy as freedom from society, meaning protecting information from others; other is privacy understood as dignity, meaning protection of intimate relationships and public reputation.[43]

Similarly, Article 8 ECHR is primarily concerned with guaranteeing the right to a person's private and family life, home and correspondence.[44] The provisions of Article 8 of the ECHR read as follows: "Right to respect for private and family life" provides that "There shall be no interference by a public authority with the exercise of this right except such as is in accordance with the law and is necessary in a democratic society in the interests of national security, public safety or the economic well-being of the country, for the prevention of disorder or crime, for the protection of health or morals, or for the protection of the rights and freedoms of others".

The right of freedom of movement is enshrined in Article 13 UDHR and includes the right to move freely within a country; the right to cross an international border, expressed as the right to leave any country; and the right to return to one's country.[45] Research into the legal regulation of free movement related to modern international human rights law will necessarily include emigration as the "highest form" of freedom of movement.[46] Restrictions for national security should be analysed together with other human rights and the means of surveillance in the interest of public order.

---

[41] UN Human Rights Office of the High Commissioner, The Universal Declaration of Human Rights at 70: Still Working to ensure freedom, Equality and Dignity for all, https://www.standup4humanrights.org/layout/files/30on30/UDHR70-30on30-article12-eng.pdf, (accessed 11.05.2019).

[42] Article 12 UDHR, *No one shall be subjected to arbitrary interference with his privacy, family, home or correspondence, nor to attacks upon his honor and reputation. Everyone has the right to the protection of the law against such interference or attacks.*

[43] Ibid. *supra* note Diggelmann, O., Cleis, M.N., How the Right to Privacy Became a Human Right, Human Rights Law Review, Volume 14, Issue 3, September 2014, pp. 441–458, https://doi.org/10.1093/hrlr/ngu014, (accessed 3.09.2019).

[44] Convention for the Protection of Human Rights and Fundamental Freedoms, (ETS No. 005), entered into force 3 September 1950. Rome, 4.XI.1950. Council of Europe.

[45] Article 13 (1) *Everyone has the right to freedom of movement and residence within the borders of each State.* (2) *Everyone has the right to leave any country, including his own, and to return to his country.*

[46] McAdam, J., An intellectual History of Freedom of Movement in International Law: The Right to Leave as a Personal Liberty, pp. 4, Melbourne Journal of International Law, https://law.unimelb.edu.au/__data/assets/pdf_file/0011/1686926/McAdam.pdf, (accessed 23.09.2019).

The second clause of Article 29 UDHR underlines that under certain conditions, limitations may be imposed to individual rights.[47] While it can be emphasized that the UDHR is by definition an instrument protecting human rights, Article 29 (2) seeks to place the burden of proof on anyone who seeks to limit them. Any limitation must be prescribed by the law and clearly articulated in "due recognition and respect for the rights and freedoms of others". Nonetheless, limitation of human rights can be justified based on "morality, public order and the general welfare in a democratic society". Such references are considered by the UDHR as a ground for the limitation of human rights and should be interpreted from the perspective of imposing a balance between human rights and national constraints. An analogy may be made to promoting the cause of international human rights while using the development of technology to support evidence in domestic and international tribunals.

1.1.2 *Data Protection and Privacy Rights*

Remote sensing from space could raise problems of privacy rights, especially when the expected high resolution becomes available to a wide range of commercial players. So far, questions regarding privacy and human rights have been mostly prevalent when referring to Unmanned Aerial Vehicles (UAVs). UAVs are capable of taking high resolution images and videos. Such technology is available for commercial purposes and it is more common for people to use on a daily basis or on limited areas of the ground. The need to protect privacy has been debated and basic rules are in place in regard to the use of UAVs. However, the proliferation of such technology could necessitate the need for specific rules on sharing images and videos. The imminent deployment of LEO satellite constellations could reasonably pose a high risk to privacy depending on the technical characteristics of the satellites able to collect high resolution images and videos. In addition, when discussing security, the new technology could support face recognition and identification of human targets, which again poses questions regarding human rights and privacy.[48]

---

47 *In the exercise of his rights and freedoms, everyone shall be subject only to such limitations as are determined by law solely for the purpose of securing due recognition and respect for the rights and freedoms of others and of meeting the just requirements of morality, public order and the general welfare in a democratic society.*

48 Masutti, A., UAV Operations: The European Union Legal Framework on Privacy and Data Protection, The Aviation and Space Journal, July–September 2018 Year XVII No. 3, http://www.aviationspacejournal.com/wp-content/uploads/2018/10/The-Aviation-Space-Journal-Year-XVII-July-September-2018.pdf, (accessed 23.09.2019).

Until now, UAVs have been employed predominantly by States, mainly for security purposes, however, UAV technology is becoming increasingly accessible to private undertakings and even individuals. UAVs are able to collect data much more efficiently than either satellites or manned aircraft ever could. Nevertheless, they (in particular micro drones) represent a real danger for privacy rights as they can collect high-resolution images and videos, telephone conversations, and intercept electronic communications and any other wave or signal depending on the payload installed on-board. They can also recognize faces, or even detect 'abnormal behavior' and identify

Both satellites and UAVs can be regarded as effective tools for continuous surveillance. Such technologies were previously used mostly for security purposes and managed nationally and internally by GOs. When comparing satellite services and UAVs in terms of costs, the UAVs may be reasonably regarded as more expensive to use than satellites, more limited in reaching and analysing large areas, and more easily targeted. However, satellites will be deployed in outer space, where there are no sovereign borders, and will be able to provide global coverage in high resolution videos and images in real-time.

With the advent of space commercialisation and low costs both for launching and manufacturing but also for services and finite products, the technology will become increasingly accessible to private companies or even individuals. This intrusion could be dangerous, and it is expected to become more advanced with the advent of AI processing big data from space.

From a regulatory standpoint, it should be underlined that in relation to privacy, the regulatory work is characterized by complexity and by different approaches to defining it and determining its focus. When regulating any activities that may compromise privacy, there are many facets that may become difficult to control such as the changing "degree of necessity of protecting certain aspects of the private lives of citizens". This was further explained in the context of cultural heritage, meaning that privacy is shaped by the habits and the culture in a given society, which may change over time.[49] To further explain, the impact of technology innovation and the development of surveillance technologies are centre-focused on human activities. Because of the advances and in particular the availability of such data from commercial companies that have the capability of better coverage and higher resolution, such opportunities may support the legal sector.[50] In this case there should be a specific instrument that would help in identifying where privacy should be protected from any intrusion and where it can be balanced between the need to analyse and identify human rights violations and the right to privacy. In general, the principles concerning proof of perpetration are described in the criminal codes of each country.

---

human targets. Finally, future solar-power drones will also be able to 'stay in the air forever' becoming a continuous surveillance tool in the hand of public authorities. This opens a quite dangerous scenario whereby individuals must now guard themselves from intrusions by other individuals. Consequently, on many occasions the European Commission has expressed its concern about the proliferation of surveillance tools, especially in regard to UAVs.

[49] Doldirina, C., What is privacy? in: Aranzamendi, M.S., Sandau, R., Schrogl, K.U. ESPI European Space Policy Institute, Current Legal Issues for Satellite Earth Observation, Report 25, August 2010, pp. 50–54.

[50] Purdy, R., Treaty Verification and Law Enforcement Through Satellite Earth Observation. In Aranzamendi, M.S., Sandau, R., Schrogl, K.U. ESPI European Space Policy Institute, Current Legal Issues for Satellite Earth Observation, Report 25, August 2010, pp. 9–13.

1.1.2.1 The United States of America

In the USA for example, the government controls and has multiple ways of restricting sensitive satellite data. One way is to just buy exclusive rights to images the government would not like to become public, or by issuing directives barring imaging over a given location. This method is focused on sensitive imaging for land, the so-called Land Remote Sensing Policy Act. This law forms the legal basis for commercial remote sensing. The "shutter-control" puts the government in the position to ban commercial operators from releasing images on certain locations. Buying exclusive rights is also known as "checkbook shutter control". Also, legislation bans U.S. companies from releasing high resolution images of Israel and the Occupied Territories, the Kyl-Bingman Amendment. In addition, licenses released for remote sensing could impose area imaging restrictions, based on national security and/or foreign policy concerns. Also, resolution limits may be imposed, restricting the maximum capability of the satellites. The strict regulations for export control of products from U.S. companies to other companies have also incentivized the development of other satellites. The U.S. regulations do not apply to foreign companies. Emergence of an international remote-sensing community necessitates reform of the licensing or regulatory processes for remote sensing, as satellite operators would like to expand their services, especially by the satellite constellations in LEO. The bill, the Space Commerce Free Enterprise Act of 2017 will provide a better process for licensing.[51]

In the USA, the National and Commercial Space Programs Act provides that a private company or individual that is under the jurisdiction and control of the USA may operate a private remote sensing space system only with a license and authorized by the Secretary of Commerce.[52] The U.S. Department of Commerce, through the National Oceanic and Atmospheric Administration (NOAA) licenses the operation of private remote sensing under the Land Remote Sensing Policy Act of 1992.[53] Back in 2018, the U.S. Department of Commerce published an Advance Notice of Proposed Rulemaking (ANPRM) that stated that it was considering revisions to its regulations for the licensing of private remote sensing space systems. The

[51] Scoles, S., How the Government Controls Sensitive Satellite Data. It might just buy exclusive rights to image it doesn't want out there, 2.08.2018, https://www.wired.com/story/how-the-government-controls-sensitive-satellite-data/, (accessed 1.09.2019).

[52] NOAA—National Oceanic and Atmospheric Administration, About Commercial Remote Sensing Regulatory Affairs, 20 March 2019, https://www.nesdis.noaa.gov/CRSRA/, (accessed 1.09.2019).

[53] NOAA—Department of Commerce, Licensing of Private Remote Sensing Space Systems, A Proposed Rule by NOAA on 14.05.2019, https://www.federalregister.gov/documents/2019/05/14/2019-09320/licensing-of-private-remote-sensing-space-systems, (accessed 1.09.2019).

Pursuant to Article VI of the Treaty on Principles Governing the Activities of States in the Exploration and Use of Outer Space, including the Moon and Other Celestial Bodies (Outer Space Treaty), activities of private U.S. entities in outer space require the "authorization and continuing supervision" of the United States Government. The Land Remote Sensing Policy Act of 1992, codified at 51 U.S.C. 60101 et seq. (ACT), authorizes the Secretary of Commerce (Secretary) to fulfill this responsibility for operators of private sensing space systems, by authorizing the Secretary to issue and enforce licenses for the operation of such systems. The Secretary's authority under the Act is currently delegated to the NOAA Assistant Administrator for Satellite and Information Services.

stated reason was the need to update the regulatory regime to facilitate technological advancements and new business models but also to address the "increased foreign competition". Only recently, in May 2019, the Department of Commerce published a Notice of Proposed Rulemaking (NPRM) on Licensing of Private Remote Sensing Space Systems in which it is proposing "to entirely rewrite the current regulations", last updated in 2006.[54] The most fundamental changes the Department of Commerce proposes are to:

- Create a two-category framework, where the license terms are commensurate with the risk posed by the remote sensing space system to the national security and international obligations of the United States;
- Conduct a full interagency review and consider custom license conditions only when a proposed system is novel and is in the higher risk category.

As an important element, the proposal refers to non-Earth imaging (remote sensing missions not in Earth orbit). The proposal refers to LEO constellations by proposing that an applicant applies for a single license to operate the constellation for remote sensing purposes.

> Commerce notes that the definition of "remote sensing space system" in the proposed rule makes clear that a license may authorize a system comprising one or more remote sensing instruments and spacecraft. By not limiting how many remote sensing instruments qualify as a system, the proposed rule permits an applicant to apply for a single license to operate a series or constellation of remote sensing instruments. So long as the characteristics and capabilities of the entire system are fully and accurately described in the application, a system comprising multiple instruments could potentially receive a single license.[55]

In addition, the proposal refers to high-risk applications "one important standard high-risk condition addresses data protection" which refers to unauthorized system access.[56] Privacy is not addressed in the proposal.

1.1.2.2 The European Union

The users of unmanned aerial systems (UAS) or UAVs, the so-called drones, could be regarded as a risk to privacy to other people. The question that may be asked to UVAs operators could resonate further to satellite operators. Therefore, is the General Data Protection Regulation (GDPR) of little interest to drone operators, or does the GDPR somehow affect the potential use of satellite constellations to use global, real-time, high resolution imagery?

---

[54]Department of Space Commerce, Commerce Proposes Streamlined Licensing Rules for Private Remote Sensing Satellites, 14.05.2019, https://www.space.commerce.gov/commerce-proposes-streamlined-licensing-rules-for-private-remote-sensing-satellites/, (accessed 1.09.2019).

[55]Subpart A: General, https://www.federalregister.gov/documents/2019/05/14/2019-09320/licensing-of-private-remote-sensing-space-systems, (accessed 23.09.2019).

[56]Subpart E: High Risk Category, https://www.federalregister.gov/documents/2019/05/14/2019-09320/licensing-of-private-remote-sensing-space-systems, (accessed 1.09.2019).

Obviously, more controversial are the UAVs as many individuals have started using drones in a public space. Such users come under GDPR jurisdiction if they are collecting personal information that could potentially be used to identify an individual. Such data include face recognition or details that would allow an individual to be identified, for example visible address, car number plates, clothing etc. Also mentioned were details of bodily characteristics, for example tattoos. Details of an individual's private or professional life or any data collected that can be used to evaluate a person, including for security or monitoring purposes, in particular any footage that tracks a person over a prolonged period of time, increases the likelihood of being identified. As a precautionary solution, there is the possibility of informing the public and referring them to the public privacy statement—a public statement posted on the website containing the way the information is gathered, used, and disclosed. The information should be anonymized, blurring information that may be used to identify a person. Also, the public has the right to access their data, receive a copy and request changes.

The solutions exist to limit the intrusion in a person's private life, but are they applied? At least theoretically yes, but what if such information, after being collected has to be shared? And most importantly, what if the information must be shared internationally? Are there any rules that apply globally? Complex matters regarding data sharing imply that data must not be shared with third parties without the explicit permission of the individual concerned, unless it is anonymized beforehand.[57]

The use of UAVs collecting personal identification information must comply with the fundamental rights to privacy and data protection. The Directorate-General for Internal Policies, Policy Department, Citizen's Rights and Constitutional Affairs of the European Parliament on Civil Liberties, Justice and Home Affairs (EU LIBE) Committee provided an in-depth analysis of the applicable European, EU and national law:

- Article 8 of the Council of Europe Convention on Human Rights and related jurisprudence
- The ECHR and related jurisprudence guarantees apply to any drones' use in relation to privacy and the related data protection implications
- Council of Europe Convention 108, Recommendation R(87)15 of the Committee of Ministers of the Council of Europe on the use of personal data in the police sector, Recommendation CM/Rec(2010)13 of the Committee of Ministers to member states on the protection of individuals with regard to automatic processing of personal data in the context of profiling
- Article 7 (privacy) and 8 (data protection) of the Charter on Fundamental Rights of the EU
- Directive 95/46/EC of the European Parliament and of the Council of 24 October 1995 on the protection of individuals with regard to the processing of personal data and on the free movement of such data
- Council Framework Decision 2008/977/JHA of 27 November 2008 on the protection of personal data processed in the framework of police and judicial cooperation in criminal matters

[57] Drones and GDPR, Rules you should follow, 3 September 2018, https://www.coptrz.com/drones-gdpr-the-rules-you-should-follow/, (accessed 27.08.2019).

- Regulation (EC) 45/2001 of 18 December 2000 on the protection of individuals with regard to the processing of personal data by the Community institutions and bodies and on the free movement of such data
- Directive 2002/58/EC of 12 July 2002 concerning the processing of personal data and the protection of privacy in the electronic communications sector (Directive on privacy and electronic communications)
- National Constitutions, laws and regulations implementing EU law, notably the Data Protection Directive, which apply as far as they cover drones activities and on the basis of national jurisprudence.

It was however noted that law enforcement activities, using the example of the police or other law enforcement bodies that process and collect data at national level, fall outside the EU laws, except where such data are exchanged between member states, when Framework Decision 2008/977/JHA applies. However, national laws must serve a legitimate role, following the ECHR and related jurisprudence. In addition, it should be noted that Article 4(2) Treaty of the European Union (TEU) provides that "national security remains the sole responsibility of each Member State". For this reason, the activities of the intelligence services fall outside the EU competences, even when they imply collection of data by UAVs or potentially by satellites. Such activities would be however regulated nationally and would have to comply with the proportionality and necessity principles at EU level.

The EU Legal Framework on Privacy and Data Protection could be used to better understand and address the privacy and data protection problems that may arise in using drones and satellites.[58]

> The privacy and data protection problems which arise from technological progress (like drones) could usefully be addressed through the application of the European Union (EU) legal framework. In fact, the EU is committed to accede to the European Convention of Human Rights (ECHR) on the basis of Article 6 of the Treaty. Article 7 of the EU Charter of Fundamental Rights (which became binding in December 2009 when the Lisbon Treaty came into force) replaced the right to privacy of 'correspondence' with the right to privacy of 'communications'. In addition, Article 8 of the Charter states that "everyone has the right to the protection of personal data concerning him or her" and that such data must be processed for specific purposes only and exclusively with the consent of the person concerned. The same provision establishes the right to access the data collected concerning oneself and, if needed, the right to have it rectified. Compliance with these rules should be subject to the control of an independent overseeing authority. Article 7 of the Charter mainly derives from Article 286 of the Treaty establishing the European Community and Directive 95/46/EC of the European Parliament and of the Council on the protection of individuals with regard to the processing of personal data and on the free movement of such data, as well as, on Article 8 of the ECHR and on the Council of Europe Convention of 28 January 1981 for the Protection of Individuals with regard to Automatic Processing of Personal Data, which has been ratified by all the Member States.[59]

[58]Ibid. *supra* note Masutti, A., UAV Operations: The European Union Legal Framework on Privacy and Data Protection, The Aviation and Space Journal, July–September 2018 Year XVII No. 3, http://www.aviationspacejournal.com/wp-content/uploads/2018/10/The-Aviation-Space-Journal-Year-XVII-July-September-2018.pdf, (accessed 23.09.2019).

[59]Ibid. *supra* note Masutti, A., UAV Operations: The European Union Legal Framework on Privacy and Data Protection, The Aviation and Space Journal, July–September 2018 Year

It is pertinent to mention that the current Data Protection Framework could be applicable to drones and satellites and would be applicable to international data sharing inside and outside the EU. Directive 95/46/EC referring to Data Protection was replaced by the GDPR and presents a number of principles for data protection and privacy:

> transparency: the data collector must notify the 'data subject' of the personal information collected, the purpose of that collection and use of the data; ii) data minimisation: data must be 'relevant' to the purpose for which it is being collected and the data collected must be the minimum amount of data necessary for the purposes pursued; iii) consent: individuals must give consent to their data being collected; iv) accountability: the data controller must be identifiable and accountable to individuals and regulatory authorities; v) right of access, correction and erasure: individuals retain control over the information that is collected about them.[60]

To underline, several privacy issues may arise in relation to the processing of data collected by remote sensing, whether UAVs or satellites:

> Such risks can range from a lack of transparency due to the difficulty of being able to view small and micro drones from the ground, to a difficulty in knowing the purposes for which the personal data is being collected.[61]

This is the reason why member states such as UK, France, Italy, Belgium, Germany and many more have already issued regulations for UAVs establishing norms on data protection and privacy rights. In fact, the privacy law concerning the telecommunication and Internet services sector in the mentioned EU member states, could be all very similar as they transcribe the EU Data Protection Directive and e-Privacy Directive:

> lack of provisions specific to UAV in regard to the collection and distribution of data and images. However, by way of analogy, the existing regulatory framework may be applicable to the use of drones, and the existing case law on data collection and handling may provide guidance in the drafting and implementation of regulation specific to drones, if necessary. In fact, the existing privacy and data protection rules are themselves sufficient to distinguish between lawful and unlawful use of drones, however without effective oversight by authorities, effective enforcement procedures, and sufficient resources and manpower, these rules are bound to remain unheeded. Moreover, privacy rights can be protected also by 'embedding' privacy laws in the technology that now threatens them. Privacy by design might actually prove to be an extremely useful tool to ensure the enforcement of EU and national legislation. This objective has been facilitated by the capacities of computers to draw upon the tools of artificial intelligence (AI) and operational research. The importance of privacy

---

XVII No. 3, http://www.aviationspacejournal.com/wp-content/uploads/2018/10/The-Aviation-Space-Journal-Year-XVII-July-September-2018.pdf, (accessed 23.09.2019).

[60] Ibid. *supra* note Masutti, A., UAV Operations: The European Union Legal Framework on Privacy and Data Protection, The Aviation and Space Journal, July–September 2018 Year XVII No. 3, http://www.aviationspacejournal.com/wp-content/uploads/2018/10/The-Aviation-Space-Journal-Year-XVII-July-September-2018.pdf, (accessed 23.09.2019).

[61] Ibid. *supra* note Masutti, A., UAV Operations: The European Union Legal Framework on Privacy and Data Protection, The Aviation and Space Journal, July–September 2018 Year XVII No. 3, http://www.aviationspacejournal.com/wp-content/uploads/2018/10/The-Aviation-Space-Journal-Year-XVII-July-September-2018.pdf, (accessed 23.09.2019).

by design had already been acknowledged by the EU lawmakers in the recent Regulation 679/2016/EU.[62]

1.1.2.3 Germany

The German Air Traffic Act addresses and defines unmanned aerial systems (UAS). In particular, the operation of a UAS that weighs more than 5 kilograms requires authorization from the aviation authority of the German state. Particular emphasis is put on data privacy and data protection rules. Authorization will be issued under the Air Traffic Regulation if the intended operations will not violate the principles of data protection and privacy rights. Also, if the UAS weighs more than 25 kg or is operated outside the visual line of sight of the operator, it is generally prohibited. It was stated on the official website that the German Ministry of Transportation and Digital Infrastructure is currently working to update the existing drone regulation regime.[63] Referring to the Federal Data Protection Act, it was mentioned that if the UAS is used to process personal data then the requirements of the Federal Data Protection Act should be respected:

> If a UAV is equipped with a video camera, the requirements of section 6b of the Federal Data Protection Act (video surveillance of public places) must be taken into account. Video surveillance of public places may only be conducted to fulfil public tasks, to exercise the right to determine who shall be allowed or denied access to a property, or to pursue rightful interests for precisely defined purposes—for example, protection against theft or vandalism. If a UAS is merely used for recreational purposes and the surveillance is not of a lasting nature, the provision will not be applicable.
>
> If the UAV camera is surveilling a non-public area, section 4, paragraph 1 of the Federal Data Protection Act applies. In such a case, any collection, processing, and use of personal data is only admissible if permitted by law or if the person has consented. Furthermore, the subjects of the surveillance have to be informed of the identity of the data collector, purposes of collection, processing, or use of their personal data, and of possible recipients.

In addition, several provisions refer to the right to control the use of one's image. In this regard the provisions of Section 22, Paragraph 1 of the Copyright Arts Domain Act apply, namely:

> images can only be disseminated with the express consent of the person concerned. Exceptions are listed in section 23—for example, for images that portray an aspect of contemporary society—on condition that the publication does not interfere with a legitimate interest of the person concerned. Dissemination includes public and private dissemination—for example, making an image available to a limited audience on the Internet.[64]

---

[62] Ibid. *supra* note Masutti, A., UAV Operations: The European Union Legal Framework on Privacy and Data Protection, The Aviation and Space Journal, July–September 2018 Year XVII No. 3, http://www.aviationspacejournal.com/wp-content/uploads/2018/10/The-Aviation-Space-Journal-Year-XVII-July-September-2018.pdf, (accessed 23.09.2019).

[63] Gesley, J., Library of Congress, Regulation of Drones: Germany, 2016, https://www.loc.gov/law/help/regulation-of-drones/germany.php#data-protection, (accessed 1.09.2019).

[64] Ibid. *supra* note Gesley, J., Library of Congress, Regulation of Drones: Germany, 2016, https://www.loc.gov/law/help/regulation-of-drones/germany.php#data-protection, (accessed 1.09.2019).

1.1.2.4 The United Kingdom of Great Britain and Northern Ireland

In the UK for example, the users of UAS or UAVs may be regarded as a risk to privacy to other people, for the reason of "having the potential" to be covered by the Data Protection Act, if used for personal use. Instead, if using the drone for a professional purpose, there are legal obligations as a data controller, and in the UK the CCTC code refers to UAS.[65]

1.1.2.5 Other

In Carpenter v. United States (2018), Carpenter was charged with six counts of robbery and six counts of carrying a firearm during a federal crime of violence. Carpenter's phone was used to produce a location map that placed him near four of the charged robberies, at exact time of the robberies. According to the opinion of the Court, such records represented a powerful circumstantial evidence of his participation, especially if his cell phone was usually not located near the stores except during the robbery. The lack of privacy raised the question if the resulting records are entitled to the Fourth Amendment protection which protects the right against unreasonable searches and seizures. In this case, the Court held that "an individual maintains a legitimate expectation of privacy in the record of his physical movements as captured through monitoring of a cell phone's location. Like other privacy rights, this right is conditional. If the government suspects wrongdoing, it can always seek judicial approval to monitor a particular person after demonstrating reasonable grounds to believe that he or she has committed or is planning a crime. But absent a judicial order, the government should not be allowed to produce a database of our movements".[66]

Human Rights Watch also analysed over 50 satellite images recorded between January 2013 and May 2018 to assess allegations of indiscriminate bombardment, monitor the construction of military bases, quantify building demolition, verify witness testimony, and validate videos and photographs shared through social media.[67] It also reviewed open-source material shared by first responders and activists after the May 25 incendiary weapons attack in Khan Sheikhoun, Idlib. And it assessed a time series of satellite imagery and identified extensive burn scars on agricultural land along the western side of Khan Sheikhoun, consistent with the video and testimony. These burn scars measured approximately 175,000 square meters in total area and were the result of intense fires occurring within a 24-h period between the morning of

[65] ICO—Information Commissioner's Office, Drones, https://ico.org.uk/your-data-matters/drones/, (accessed 27.08.2019).

[66] Roth, K., Blindfold Big Brother: We need a Global Right to Privacy in Public Space, 16 June 2019, https://www.hrw.org/news/2019/06/16/blindfold-big-brother-we-need-global-right-privacy-public-spaces, (accessed 25.09.2019).

[67] HRW, If You Are Afraid for Your Lives, Leave Sinai! Egyptian Security Forces and ISIS-Affiliate Abuses in North Sinai, 28 May 2019, https://www.hrw.org/report/2019/05/28/if-you-are-afraid-your-lives-leave-sinai/egyptian-security-forces-and-isis, (accessed 26.09.2019).

May 24 and 25, 2019.[68] Human Rights Watch also reviewed satellite imagery showing over 70 buildings affected by fire and photographs and videos showing extensive destruction of property.[69] It also reviewed scores of official statements, social media posts, media reports, and dozens of satellite images to identify home demolitions and secret military detention facilities.[70] Human Rights Watch determined, based on in-person interviews with witnesses, videos, and satellite imagery, that the contents of the warehouse had caught fire and exploded.[71]

1.1.3 Governmental Surveillance and Privacy in the Digital Age

The Special Rapporteur on the right to privacy reports annually to the Human Rights Council, which aims to contribute to raising the level of respect, protection and fulfilment of the right to privacy.[72] The oversight mechanism is a strong basis for future policy to provide proportionate measures in a democratic society.[73] The issue of security and surveillance from space includes its own specific challenges and issues with regard to the right of privacy however, and notions such as Big Data and Open Data need to be analysed together with the issue of security, criminal offences and violations of human rights.

When referring to the right of privacy in the digital age, perhaps the most conclusive example of advancing the technology are the actual practices of national security services shared by Edward Snowden. However, the focus of state surveillance was on communications, their interception and the collection of personal data.[74] In Europe, on 21 December 2016, the Court of Justice (ECJ) delivered a judgment referring to the promotion and protection of the human right to privacy in the digital age. The ECJ judgment underlined the negative potential consequences for the exercise of

---

[68] HRW, Russia/Syria: Flurry of Prohibited Weapons Attacks, International Norms Ignored as Civilians Suffer, 3.06.2019, https://www.hrw.org/news/2019/06/03/russia/syria-flurry-prohibited-weapons-attacks, (accessed 23.09.2019).

[69] HRW, Cameroon: Promised Investigation Should be Independent. Government Forces on Rampage in North–West Region City, 23.05.2019, https://www.hrw.org/news/2019/05/23/cameroon-promised-investigation-should-be-independent, (accessed 23.09.2019).

[70] HRW, Egypt: Serious Abuses, War Crimes in North Sinai. Army's Conduct Crushed, Alienated Local Residents, 28.05.2019, https://www.hrw.org/news/2019/05/28/egypt-serious-abuses-war-crimes-north-sinai, (accessed 23.09.2019).

[71] HRW, Yemen: Warehouse Blast Kills Schoolchildren. Houthis Stored Volatile Material in Residential Area, 9.05.2019, https://www.hrw.org/news/2019/05/09/yemen-warehouse-blast-kills-schoolchildren, (accessed 23.09.2019).

[72] Resolution 28/16 of the Human Rights Council, April 2015, The right to privacy in the digital age, https://www.right-docs.org/doc/a-hrc-res-28-16/, (accessed 23.09.2019).

[73] Human Rights Council, A/HRC/34/60, Paragraph 4, Agenda item 3 Promotion and protection of all human rights, civil, political, economic, social and cultural rights, including the right to development.

[74] Ibid. *supra* note, A/HRC/34/60, Paragraph 10–11.

freedom of expression, mentioning that "the fact that the data is retained without the subscriber or registered user being informed is likely to cause the persons concerned to feel that their private lives are the subject of constant surveillance".[75]

In this context, it needs to be strongly underlined that the virtue of such judgement is to put into balance the protection of the private life with the effective oversight mechanism. The judges underlined that "While the effectiveness of the fight against serious crime, in particular organized crime and terrorism, may depend to a great extent on the use of modern investigation techniques, such an objective of general interest, however fundamental it may be, cannot in itself justify that national legislation providing for the general and indiscriminate retention of all traffic and location data should be considered to be necessary for the purposes of that fight".[76] This rigorous analysis of proportionality shows the importance and need to develop procedures to protect against interference with privacy rights.

1.2 Human Rights and Space Legislation

Space technology is by definition at the forefront of innovation and states are noting that the benefits derived from the application of space technologies play a major role in everyday life.[77] Many of the space technologies are utilized back on Earth. Different spin off companies adapt space technology and disseminate it for the masses. When referring to remote sensing, attention is focused on the advance of space technology in providing high resolution imagery, high speeds of delivering such data, and the processing capabilities of big data with AI systems on board the satellites. The fundamental rights of each individual require protection in every scenario of the future development of technology, and further discussions are necessary on how space can contribute to the protection of human rights but also to the protection of the right to privacy from digital interference. The rapid progress of remote sensing technology makes space data a very accurate and up-to-date spatial data resource, with multiple utilizations, such as cartographic and thematic maps, meteorology and imaging. For this reason, it is important to understand the potential of space technology to identify human rights violations and be used as evidence before the authorities to condemn the perpetrators.

Satellite data or satellite imagery also known as "Earth observation imagery" is images of Earth collected by imaging satellites. Remote sensing satellites are capable of high-resolution satellite images and remote sensing data require additional influencing factors than when using traditional imagery with optical-electronic scanners

---

[75] European Court of Justice, Tele 2 Sverige, C-203/15, ECLI:EU:C:2016:970, 21.12.2016, mn. 100.

[76] Ibid. *supra* note European Court of Justice, Tele 2 Sverige, C-203/15, ECLI:EU:C:2016:970, 21.12.2016, mn. 103.

[77] United Nations General Assembly, Committee on the Peaceful Uses of Outer Space, Legal Subcommittee Fifty-eight session, Vienna, 1–12 April 2019, Doc. A/AC/105/C.2/L.310, 16 January 2019.

characterized by sub-meter spatial resolution.[78] They can be government owned, quasi-private, or private and government.[79] UNOOSA referred to remote sensing by satellites as "the science of gathering data about objects or areas from a distance. It is a tool frequently used to obtain details about the Earth's surface from space, as data is gathered by detecting and measuring electromagnetic waves emitted, reflected or diffracted by the sensed object."[80] Remote sensing uses electromagnetic energy to capture data. Such technology is an indispensable element in the evolution of human society especially "in monitoring and assessing the resources of the Earth. The fact that it offers the advantage of viewing large sections of the Earth at regular intervals has been a significant factor in enhancing its awesome potential".[81]

Relevant rules are found in international law, in particular in space law. Remote sensing is the activity of gathering information without physical contact and can be made from UAVs, planes, balloons, satellites etc.[82] Remote sensing can provide data for many purposes, including military, environmental, meteorology and disaster management.[83] Gathering raw data on Earth from satellites is a particular activity that involves the use of space for sensing the Earth. The use of remote sensing could be generally linked with humanitarian purposes, with disaster management and weather forecasting. However, satellite imagery was not referred to in relation to commercial uses with the human factor included as the sensed objective. It is necessary to analyse and understand in which situations commercial/governmental use of satellite imagery may violate privacy and/or other human rights and if a balance can be found between the use of satellite imagery as evidence in courts to prove violations of human rights and detect the perpetrators, and the need to protect human rights.

Sensing can be active or passive. The difference is the source of radiation, i.e. whether it is reflected radiation (passive remote sensing) where the satellite sensors systems only receive the radiation and transmit the data on the ground. Passive sensors mean that no light or energy is sent to the target, meaning that the receiving of data is limited to the energy emitted or scattered by the target. Active sensors mean that they provide their own source of electromagnetic radiation. The satellite itself emits radiation in the direction of the target to be investigated. The sensor from the satellite

---

[78] Mozgovoy, D., Svynarenko, D., Tsarev, R., Yamskikh, T.N., Fast satellite imagery of lengthy territories with complex configuration, November 2018, https://www.researchgate.net/publication/329328833_Fast_satellite_imagery_of_lengthy_territories_with_complex_configuration, (accessed 23.09.2019).

[79] Gabrynowicz, J.I., A Brief Survey of Remote Sensing Law Around the World, at UN/Thailand Workshop on Space Law, Activities of States in Outer Space in Light of the New Developments: Meeting International Responsibilities and Establishing National Legal and Policy Frameworks, http://www.unoosa.org/pdf/pres/2010/SLW2010/02-13.pdf, (accessed 23.09.2019).

[80] UNOOSA, Remote Sensing, http://www.unoosa.org/oosa/en/ourwork/topics/remote-sensing.html, (accessed 23.09.2019).

[81] Diederiks-Verschoor, I.H.Ph., Kopal, V, An Introduction to Space Law, pp. 70, Kluwer Law International, 3rd Edition, 19 February 2008.

[82] Tronchetti, F., Fundamentals of Space Law and Policy, pp. 16, Springer Briefs in Space Development, Springer, 2013 Edition, 30 June 2013.

[83] Lyall, F., Larsen, P.B., Space Law: A Treatise. pp. 411–412, Routledge, 2nd Edition, 19 December 2017.

will than detect and measure the radiation that is reflected or backscattered from the target.[84]

There are three methods of acquiring information by a remote sensor: (i) reflection; (ii) emission; (iii) emission-reflection. Satellite imagery can be: (i) marine; (ii) terrestrial; (iii) atmospheric. The satellite imagery generally falls into two categories: multispectral or hyperspectral imagery. The main difference between the two is the number of (wavelength or frequency) bands within the electromagnetic spectrum that are observed and how narrow the bands are.[85] Satellite imagery can be low or high resolution, current or archived, and coming from different types of sensors. Compared with the 30+ m resolution available in the 1980s, currently possibility access to resolutions of 2.5 m or even less is available.[86] If the resolution is high or very high, more restrictions are likely to exist. These could be about national security, privacy (human related) and/or cost related. Most of the high resolution data come from commercial satellites.[87] Low resolution comes from governmental satellites with an open access policy and should be easy to access and collect. Satellite imagery is available commercially or in different archives, including the Internet, and is regarded as a means of gathering large amounts of multi-layered data that may be synthesized for a deeper understanding of different landscapes and regions. A satellite image has the potential to become less expensive and replace aerial photography in given services while supporting in situ analysis.[88]

The first aspect that has to be analysed in connection to remote sensing is the relation between the state operating a remote sensing satellite (the sensing state) and the state whose territory is sensed from space (the sensed state).[89] The provisions of the OST of 1967 and the Remote Sensing Principles of 1986 allow remote sensing from space with the following conditions: (i) the sensing state has the right to observe another state's territory from space without prior authorization; (ii) the sensed state shall have access to "primary" and "processed data" but also "analysed information" of its territory "on a non-discriminatory basis and on reasonable cost terms"; the sensed state will have access to technical assistance in order to make use of the shared data, based on mutually agreed terms.[90]

---

[84] NASA Earth Data, Remote Sensors, 3 September 2019, https://earthdata.nasa.gov/learn/remote-sensors, (accessed 26.09.2019).

[85] Hebden K., First European satellite with AI set for launch, 16.09.2019, https://room.eu.com/news/first-european-satellite-with-ai-set-for-launch, (accessed 20.09.2019).

[86] Ibid. *supra* note Lyall, F., Larsen, P.B., Space Law: A Treatise, 2nd Edition, pp. 425.

[87] Schoenmaker, A., Community Remote Sensing Legal Issues, April 2011, pp. 3 https://swfound.org/media/62081/schoenmaker_paper_community_remote_sensing_legal_issues_final.pdf, (accessed 23.09.2019).

[88] Ireland T, Urwin N, Satellite Imagery and Landscape Archaeology: An Interim Report on the Environmental Component of the Vinhais Landscape Archaeology Project, North Portugal, https://www.jstor.org/stable/24667823?seq=1#page_scan_tab_contents, (accessed 23.09.2019).

[89] Ibid. *supra* note Tronchetti, F., Fundamentals of Space Law and Policy, pp. 16, Springer Briefs in Space Development, Springer, 2013 Edition, 30 June 2013.

[90] Ibid. *supra* note Tronchetti, F., Fundamentals of Space Law and Policy, pp. 16, Springer Briefs in Space Development, Springer, 2013 Edition, 30 June 2013.

The second aspect refers to the sharing and selling of remote sensing data. Such information is regulated by "data policies" that contain rules on copyrights, marketability and use of remote sensed products.[91] The main legal concerns are: (i) privacy; (ii) intellectual property; (iii) liability and (iv) national security.[92] Steps involve: (i) data collection; (ii) data access and usage; and (iii) sharing and distribution of data and/or information. SRS comprises the "space segment" and the "ground segment".[93] "Space Segment functions include the collecting, recording and transmitting to Earth of data concerning the Earth's surface. In the ground segment, the reception, conversion and interpretation of the data take place".[94] The final product is then distributed to the customers and this is the function that more often gave rise to controversy. Distribution has become a major point of issue.[95]

The availability and cost of data is one of the most important factors in the selection of platforms providing remote sensing data, an area in satellite imagery presents a series of advantages.[96] First of all, alternatives are *in situ* observations or aerial photography. For aerial images, their collection might be difficult especially when referring to sovereignty and/or the risks of surveilling areas in conflict zones. Heavy restrictions exist in some countries that deny authorization for aerial photography of their territory and these images are usually more costly to gather.[97] Two factors are considered major challenges, (i) the reliability and independence of the information; (ii) its accessibility.[98] Accessibility refers to price on the one hand but also on the protection of human rights. Reliability and independence refer to the processing of the information and the need for certification to be used in courts. Combining ground-based aerial and space-based observations is a way to generate more precise information.[99] Generally, the remote sensing is considered to lead to "improved

[91] Ibid. *supra* note Tronchetti, F., Fundamentals of Space Law and Policy, pp. 16, Springer Briefs in Space Development, Springer, 2013 Edition, 30 June 2013.

[92] Ibid. *supra* note Schoenmaker, A., Community Remote Sensing Legal Issues, April 2011, https://swfound.org/media/62081/schoenmaker_paper_community_remote_sensing_legal_issues_final.pdf, (accessed 23.09.2019).

[93] Ibid. *supra* note Diederiks-Verschoor, I.H.Ph., Kopal, V, An Introduction to Space Law, pp. 71, Kluwer Law International, 3rd Edition, 19 February 2008.

[94] Ibid. *supra* note Diederiks-Verschoor, I.H.Ph., Kopal, V, An Introduction to Space Law, pp. 71, Kluwer Law International, 3rd Edition, 19 February 2008.

[95] Ibid. supra note Diederiks-Verschoor, I.H.Ph., Kopal, V, An Introduction to Space Law, pp. 71, Kluwer Law International, 3rd Edition, 19 February 2008.

[96] Gonzalez, F.E., Ruiz, M.J., Acosta, F.M., Remote Sensing Tutorial Telecan Program for the Development of Technological Networks and Application of Remote Sensing Data in West Africa, https://www.grss-ieee.org/wp-content/uploads/2014/07/EN_TUTORIAL_COMPLETO.pdf, (accessed 23.09.2019).

[97] Ibid. *supra* note bid. *supra* note Schoenmaker, A., Community Remote Sensing Legal Issues, April 2011, pp. 3, https://swfound.org/media/62081/schoenmaker_paper_community_remote_sensing_legal_issues_final.pdf, (accessed 23.09.2019).

[98] Ibid. *supra* note Diederiks-Verschoor, I.H.Ph., Kopal, V, An Introduction to Space Law, pp. 71, Kluwer Law International, 3rd Edition, 19 February 2008.

[99] A/AC.105/2015/CRP.9, The United Nations/Germany International Conference on Earth Observation—Global Solutions for the Challenges of Sustainable Development in Societies

quality of the information provided to decision makers as a way to facilitate the decision making process" including international negotiations. Aerial photography achieves a higher spatial resolution, and more objects can be detected. Satellites have a higher temporal resolution as they can gather data over a longer period of time. Also, satellites have better coverage and can see wider areas.

Traditional "film" systems differ from satellite data, which are digitalized data. Film systems involve the capture of data by the alteration of chemical layers on the film while the digital systems capture the wavelengths of radiation reaching the individual pixels of a sensor array with the advantage that that data can be transmitted by radio to a ground station, and then that data can be processed by appropriate computer programs.[100] Remote sensing from space at first depended on the use of orthodox film, canisters of film of Earth objects being ejected from satellites and recovered for processing. The development of remote digital technology currently enables the transmission of data from satellites to ground stations for later analysis and interpretation.[101]

Taking into consideration the legality of sensing operations and regarding them as lawful as regards the data collected, then the core of the analysis should take into consideration the aspects of authorization of the data, the security aspects of the content. Remote sensing by satellites has developed and is no longer exclusive for states but is also used by commercial entities, thus creating a sort of a mixed economy. Some of the companies handle and interpret data provided by state systems.[102] Satellite Remote Sensing (SRS) has become a powerful aid in monitoring and assessing the resources of the Earth.[103] Earth-orbiting satellites have enabled major progress in the field of meteorology. The definition of remote sensing does not contain the human factor. What is the relation to human rights?

The OST[104] provides for the freedom of access to outer space. Also, the 1986 UN Principles relating to remote sensing of the Earth from space provide for the freedom of remotely sensing the Earth and collecting information. Both instruments call for international cooperation and dissemination of the information, on particular the collected data to all the sensed states that might benefit from it. A further analysis will underline the points of interest for remote sensing and the relation of space law with international law.

---

at Risk, 2015, http://www.unoosa.org/oosa/oosadoc/data/documents/2015/aac.1052015crp/aac.1052015crp.9_0.html, (accessed 23.09.2019).

[100] Ibid. *supra* note Lyall, F., Larsen, P.B., Space Law: A Treatise, 2nd Edition, pp. 413.

[101] Ibid. *supra* note Lyall, F., Larsen, P.B., Space Law: A Treatise, 2nd Edition, pp. 413.

[102] Ibid. *supra* note Lyall, F., Larsen, P.B., Space Law: A Treatise, 2nd Edition, pp. 414.

[103] Ibid. *supra* note Diederiks-Verschoor, I.H.Ph., Kopal, V, An Introduction to Space Law, pp. 70, Kluwer Law International, 3rd Edition, 19 February 2008.

[104] UNGA RES 2222 (XXI) 1966, The UN Treaty on Principles Governing the Activities of States in the Exploration and Use of Outer Space, including the Moon and Other Celestial Bodies, of 27 January 1967, 8 UST 2410, 610 UNTS 205, 6 ILM 38, 1967.

1.2.1 *Outer Space Treaty and Human Rights Implications and Considerations*

International public law examines the conduct of states among themselves as well as the conduct of states towards individuals, and encompasses legal instruments such as the UN system of treaties, and bilateral or multilateral intergovernmental treaties or organizations, including other non-binding instruments.[105] The future prospects of international law are necessarily interrelated with advancements in technology. Space is a dynamic environment and space law is struggling to adapt and allow innovative activities in outer space to support the peaceful uses of outer space. Space law is part of international space law and incorporates both international and domestic instruments referring to the activities in outer space and has been described as a "body of law perhaps more than any other body of law moving (hopefully) forward" calling for adaptation.[106]

The Magna Carta of Space Law is the Treaty on Principles Governing the Activities of States in the Exploration and Use of Outer Space, including the Moon and Other Celestial Bodies.[107] With 109 ratifications, the OST is the most successful instrument of space law.[108]

The OST of 1967 provides the basic framework on international space law, and is the treaty containing principles a limited number of which could be reasonably referred to as "the accepted body of customary law principles regarding space". These include: (i) the free use of space by all states; (ii) a prohibition on claims of sovereignty over space; (iii) free exploration; (iv) the obligation to rescue astronauts in distress.[109] However, the debate is extended and complex when determining which principles are binding as matters of both customary law and *lex scripta* and this problem would warrant a paper on its own. The 50th anniversary of the OST has been recently celebrated. The legal regime of outer space has been created with leading principles

[105] White, H.M. Jr., International Law and Relations, https://er.jsc.nasa.gov/seh/law.html, (accessed 23.09.2019).

[106] Von der Dunk, F., United Nations Principles on Remote Sensing and the User, 2002, Digital Commons University of Nebraska, https://digitalcommons.unl.edu/cgi/viewcontent.cgi?article=1016&context=spacelaw, (accessed 23.09.2019).

[107] UNGA RES 2222 (XXI) 1966, The UN Treaty on Principles Governing the Activities of States in the Exploration and Use of Outer Space, including the Moon and Other Celestial Bodies, of 27 January 1967, 8 UST 2410, 610 UNTS 205, 6 ILM 38, 1967.

[108] The treaty is based on three General Assembly Resolutions: UNGA A/RES/1802 (XVII) of 14 December 1962—International Cooperation in the Peaceful Uses of Outer Space; UNGA A/RES/1962 (XVIII) of 13 December 1963—Declaration of Legal Principles Governing Activities of States in the Exploration and Use of Outer Space; UNGA A/RES/1963 (XVIII) of 13 December 1963—International Cooperation in the Peaceful Uses of Outer Space. See also: A/AC.105/C.2/2019/CRP.3*, Status of International Agreements relating to activities in outer space as at 1 January 2019, 1 April 2019, http://www.unoosa.org/documents/pdf/spacelaw/treatystatus/AC105_C2_2019_CRP03E.pdf, (accessed 23.09.2019).

[109] Schmitt, M.N., International Law and Military Operations in Space, pp. 99, Max Planck Yearbook of United Nations Law, Volume 10, 2006, pp. 89–125, https://www.mpil.de/files/pdf3/04_schmittii1.pdf, (accessed 23.09.2019).

contained in the OST, which were skilfully summarized by Diederiks-Verschoor and Kopal in 2008.[110] While the OST is based on principles containing only rules of a general nature, the remaining four treaties include more detailed provisions.[111]

The OST incorporates rules of general international law, however, a large number, although not expressly mentioned in OST, are *ipso jure* to be applied to outer space.[112] The legal regime of outer space has many similarities with the legal regime of the high seas, such as its non-appropriative character. U.S. President John F. Kennedy when addressing a US University in 1962, referred to outer space as the "new sea", "new ocean" or "sea of peace" where new rights have to be won and used for the progress of all people, while emphasizing the peaceful purposes of outer space.[113] The variety of rules in the OST must be noted. Overall, the OST declares rules of general international law such as Article III, it applies existing rules to state activities such as the provisions of Article I and Article II, and it has norm creating provisions such as Article I paragraph 1. Human rights in conjunction with these principles can be further analysed with reference to remote sensing activities.

Article III of the OST is inclusive about the relation of space law to international law: "States Parties to the Treaty shall carry on activities in the exploration and use

---

110 Diederiks-Verschoor, I.H.Ph., Kopal, V, An Introduction to Space Law, pp. 83, Kluwer Law International, 3rd Edition, 19 February 2008:

"(1) The exploration and use of outer space, including the Moon and other celestial bodies, shall be carried out for the benefit and in the interest of all countries.

(2) Outer space shall be free for exploration and use by all states on a basis of equality.

(3) Outer space shall not be subject to appropriation by claim of sovereignty, by means of use or occupation, or by any other means.

(4) Activities in the exploration and use of outer space must be carried out in accordance with international law, including the Charter of the United Nations, in the interest of maintaining international peace and security.

(5) No nuclear weapons or any other kinds of weapons of mass destruction shall be allowed to be placed in orbit around the Earth.

(6) The Moon and other celestial bodies shall be used by all States Parties to the Treaty exclusively for peaceful purposes.

(7) International cooperation and understanding are to be promoted.

(8) Astronauts shall be given every possible assistance.

(9) States Parties bear international responsibility for national activities in outer space.

(10) States Parties on whose registries the space objects are carried keep jurisdiction and control over such objects and the personnel thereof recorded in their registries.

(11) Consultations must take place in the event of dangerous activities in space. The UN Secretary-General must be informed about space activities, information which he should disseminate.

(12) All stations, installations etc. shall be open to representatives of other States Parties on a basis of reciprocity."

111 Ibid. *supra* note Tronchetti, F., Fundamentals of Space Law and Policy, pp. 8, Springer Briefs in Space Development, Springer, 2013 Edition, 30 June 2013.

112 Cepelka, C., Gilmour, J.H.C. The application of general international law in outer space, in Journal of Air Law and Commerce, Vol. 36, Issue 1, 1970, https://scholar.smu.edu/cgi/viewcontent.cgi?article=2674&context=jalc, (accessed 8 May, 2019).

113 Ibid. *supra* note Schmitt, M.N., International Law and Military Operations in Space, pp. 89–90, Max Planck Yearbook of United Nations Law, Volume 10, 2006, pp. 89–125. https://www.mpil.de/files/pdf3/04_schmittii1.pdf, (accessed 23.09.2019).

of outer space, (…), in accordance with international law, including the Charter of the United Nations, in the interest of maintaining international peace and security and promoting international cooperation and understanding".

Certain duties exist in the field of human rights as well, specifically focusing on the rules that offer personal jurisdiction to nationals abroad and protect them against violations of international law.[114] In particular, the provisions of OST Article I provide that outer space is to be used "for the benefit and in the interests of all countries (…) without discrimination of any kind, on a basis of equality and in accordance with international law". The use of satellites for remote sensing including surveillance by governments or private entities for different state or commercial purposes (for defense and intelligence related activities for national security or other purposes) is considered authorized under the provisions of OST Article I, requiring no authorization from the sensed state.

The provisions of OST Article III: "State Parties to the Treaty shall carry on activities (…) in accordance with international law, including the UN Charter in the interest of maintaining international peace and security (…)" prohibits activities in violation of international law. All obligations under the OST must be interpreted in light of the UN Charter.[115] The provisions of OST Article V provide that State Parties to the Treaty shall regard astronauts as "envoys of mankind" and provide to them "all possible assistance". Such help will be offered both on Earth in the event of an accident, distress, or emergency landing on the territory of another State Party or on the high seas, but also in outer space. Furthermore, such assistance could include, but is not limited to, access to all stations, installations, equipment and space vehicles deployed on the surface or in immediate vicinity of the celestial bodies. In this case, the provisions of OST Article XII provide that such facilities shall be open "to the representatives of other States Parties" on a basis of reciprocity. The analysis of the lack of reciprocity and the level of access to foreign facilities in outer space is however to be assessed according to individual cases, taking into consideration human rights and keeping in mind that a state cannot act in violation of international law. Another level of complexity is added when evaluating the protection of human rights if the facility pertains to a private company.

OST Article I establishes that space exploration should be used for the benefit and in the interest of all, including developing countries. The scope of Article I OST is very broad and gives room to lengthy debates on the complex meaning and the nature of international co-operation between states in outer space, and the relationship between developing and developed countries, in particular the nature of the obligation placed on those states that conduct space activities and are party to the

---

[114] Ibid. *supra* note Cepelka, C., Gilmour, J.H.C. The application of general international law in outer space, in Journal of Air Law and Commerce, Vol. 36, Issue 1, 1970, pp. 36.

[115] Ibid. *supra* note Schmitt, M.N., International Law and Military Operations in Space, pp. 102, Max Planck Yearbook of United Nations Law, Volume 10, 2006, pp. 89–125, https://www.mpil.de/files/pdf3/04_schmittii1.pdf, (accessed 23.09.2019).

OST.[116] As space is to be used for the benefit of all, remote sensing could provide benefits through the interpretation of the data from satellites, including appropriate action in case of disaster management and humanitarian assistance.

1.2.2 *UN Principles on Remote Sensing and Human Rights Implications and Considerations*

The "Principles Relating to Remote Sensing of the Earth from Outer Space" were adopted in 1986 as a General Assembly Resolution.[117] The COPUOS draft was presented to the General Assembly and duly adopted without vote on 3 December 1986 as UNGA Res. 41/65. The Principles Relating to Remote Sensing of the Earth from Outer Space are 15 in number and their adoption was considered as a means to strengthen international cooperation in this field. The Principles do not cover all remote sensing and are not exhaustive of the potentialities of the techniques.[118] The definition of"remote sensing" in Principle 1 is "the sensing of the Earth's surface from space by making use of the properties of electromagnetic waves emitted, reflected or diffracted by the sensed objects, for the purpose of improving natural resources management, land use and the protection of the environment". If remote sensing is used for other purposes, such use is not subject to the 1986 UN Remote Sensing Principles Regime since the scope is limited to "natural resources and environmental management".[119]

For the purposes of these principles with respect to remote sensing activities, Principle I can be regarded as a restriction of its application "for the purposes of improving natural resources management, land use and the protection of the environment". The subsequent paragraphs make a distinction between "primary data", "processed data" and "analysed information". The categorization of remote sensing activities in three categories of remote sensing data is useful for understanding the nature and interaction of the data with the human rights. Diverse techniques of processing are necessary for the availability of the data. Satellites scan the Earth and only at a later stage is the image processed, and when made available it must comply with the principles. However, as noted above, since such usage would not require better resolution than 10 metres, the aspects of Very High Resolution (VHR) might

[116]Jasentuliyana, N., Article I of the Outer Space Treaty Revisited, 17 Journal of Space Law, Hein Online, 1989, https://heinonline.org/HOL/LandingPage?handle=hein.journals/jrlsl17&div=16&id=&page=, (accessed 23.09.2019).

[117]'Principles Relating to the Remote Sensing of the Earth from Outer Space', 3 December 1986; UNGA Res. 41/65. cf. (1986) 25 ILM 1334–6 with note at 1331.

[118]Ibid. *supra* note Lyall, F., Larsen, P.B., Space Law: A Treatise 2nd Edition, pp. 421.

[119]Ibid. *supra* note Diederiks-Verschoor, I.H.Ph., Kopal, V, An Introduction to Space Law, pp. 83, Kluwer Law International, 3rd Edition, 19 February, 2008. *See also*: Hoskova, M., Legal Aspects of Using Remote Sensing Systems as National Technical Means of Verification, Proceedings 41st Colloquium (Melbourne, 1998), pp. 89–96.

fall outside of this Resolution. Also, the privacy aspects of VHR remote sensing data that could be disseminated commercially have not been considered.[120]

Principle II reaffirms the principles from the OST by quoting that remote sensing activities "shall be carried out for the benefit and in the interests of all countries, irrespective of their degree of economic, social or scientific and technological development". The very general reference to the benefit and interest of all countries underlining the needs of the developing countries should be given particular consideration, as it was regarded as being of considerable preponderance in international space law treaties and resolutions. This is why it should be also considered that UNGA Res. 51/122, of 13 December 1996 left complete freedom to the states to "determine all aspects" of such cooperation, repeatedly referring to the requirement of "an all equitable and mutually acceptable basis" for any activities undertaken in its implementation.[121]

Principle VIII affirms the role of the UN, proclaiming that the UN system shall promote international cooperation. This includes technical assistance and coordination. If related to the provisions of Article XI OST, state parties conducting remote sensing activities shall notify the Secretary General. If correlated with Principle XIII of the Resolution, this condition does not impose prior authorization or consent for remote sensing activities as the text of the article refers to prior "consultation" and not "prior consent".[122] However, these consultations should not necessarily be entered into before the beginning of the remote sensing activity. The scope of the consultation is to promote and intensify international cooperation. The reference to "mutually agreed terms" means no specific measure was considered for sharing the technical knowledge. Principle XII affirms the sensed state shall have access to the primary and processed data on a non-discriminatory basis and at reasonable costs, saying that a state without the capability of having satellites in orbit will have access to the processed data on the basis of reasonable costs. It also means that the sensed state has no right of veto to prevent it from being sensed, and access will be made available on "a non-discriminatory basis and on reasonable cost terms".[123]

The catalyst for change is the privatization of remote sensing. The continuous development of EO technology, such as sensors offering higher resolution, provides

---

[120] Ibid. *supra* note *Von* der Dunk, F., United Nations Principles on Remote Sensing and the User, 2002, Digital Commons University of Nebraska, https://digitalcommons.unl.edu/cgi/viewcontent.cgi?article=1016&context=spacelaw, (accessed 23.09.2019).

[121] Ibid. *supra* note Von der Dunk, F., United Nations Principles on Remote Sensing and the User, 2002, Digital Commons University of Nebraska, https://digitalcommons.unl.edu/cgi/viewcontent.cgi?article=1016&context=spacelaw, (accessed 23.09.2019).

[122] Ibid. *supra* note Diederiks-Verschoor, I.H.Ph., Kopal, V, An Introduction to Space Law, pp. 80, Kluwer Law International, 3rd Edition, 19 February 2008.

[123] Ibid. *supra* note Von der Dunk, F., United Nations Principles on Remote Sensing and the User, 2002, Digital Commons University of Nebraska, https://digitalcommons.unl.edu/cgi/viewcontent.cgi?article=1016&context=spacelaw, (accessed 23.09.2019).

new business plans and new possibilities for independent documentation of humanitarian crises.[124] Technical training in data handling is not sufficient in the new space economy to guarantee the probatory value of satellite remote sensing. As noted, in 1978, a proposal to create an International Satellite Monitoring Agency (ISMA) was tabled by France during the first session of the Disarmament Conference of the UN held in New York.[125] This proposal included the acquisition, processing and dissemination of data and/or information. Even if such organization was military designed, the relevant information included the analysed results of the data put at its disposal by satellite-operating states. The topic of technical means of verification of processed data may become very actual and a national agency or an international organization may be necessary to include the advent of commercial satellite operators generating more satellite imagery with very high resolution.

1.2.2.1 The Relationship Between Space Law and Human Rights

Various links exist between space law and human rights.[126] Article III OST is inclusive about the relation of space law with international law: "States Parties to the Treaty shall carry on activities in the exploration and use of outer space, (...), in accordance with international law, including the Charter of the United Nations, in the interest of maintaining international peace and security and promoting international cooperation and understanding". Article III OST states that international law principles are applicable in outer space. Reference to the UN Charter human rights is also included in this provision.

Space law addresses advanced, state-of-the-art technology that is constantly evolving.[127] Directly proportional with the advancement of technology, the regulatory system should also take a leap forward and provide the necessary legislation to allow the use of satellite imagery to document human rights abuses and humanitarian disasters. It was generally acknowledged at UN level that there is a collective responsibility (and not just a discretionary right) to act in the face of serious violations of human rights. It was considered imperative to protect, through prevention, by helping states to fulfil their international human rights obligations. Such obligations were developed under the Responsibility to Protect (RtoP) agreed to at the 2005 World Summit

[124] Amnesty International, Use of Satellite Imagery to document human rights abuses and humanitarian disasters. Possibilities, potentials, and putting it into action, A report conducted for Amnesty International Danish Section, http://www.prinsengineering.com/Remote%20sensing%20draft2_2005.pdf, (accessed 23.09.2019).

[125] Ibid. *supra* note Diederiks-Verschoor, I.H.Ph., Kopal, V, An Introduction to Space Law, pp. 81, Kluwer Law International, 3rd Edition, 19 February 2008.

[126] Tavernier, P., Kuskuvelis, I.I., Space Law and Human Rights: A Complementary Relationship Through the Right to Development, Published by the AIAA, first version in 11–12 May, 1990, http://www.kouskouvelis.gr/site/files/Space_Law_and_Human_Rights_A_Complementary_Relationship_Through_the_Right_to_Development.pdf, (accessed 23.09.2019).

[127] Ibid. *supra* note Tronchetti, F., Fundamentals of Space Law and Policy, pp. 8, Springer Briefs in Space Development, Springer, 2013 Edition, 30 June, 2013.

in New York for the High-level Plenary Meeting of the General Assembly.[128] Under the RtoP, respect for basic human rights principles of non-discrimination and equality, be they legal or constitutional protections, diminishes the risk that atrocity crimes are likely to occur. Former Secretary General Kofi Annan emphasized that the goal of RtoP was not to develop new law but rather to strengthen states' commitment to existing legal obligations.[129]

The performance of states in the implementation of the core international human rights treaties may be reviewed from the perspective of the implementation of remote sensing procedures. Training in remote sensing analysis should be provided to the judges, investigation teams and prosecutors of international criminal courts and tribunals, as well as the regional human rights courts and commissions. With an improved understanding among judges and other relevant court personnel of the value of remote sensing analysis, it is expected that geospatial evidence will be integrated more frequently into human rights litigation, and as a result, the work of these human rights courts and commissions will be strengthened.[130]

1.2.2.2 Remote Sensing and Digital Image Processing

Human factors play a critical role in the design and interpretation of remotely sensed imagery. In particular, a complex infrastructure is needed for the interpretation of remote sensing information. Satellite imagery is transmitted to ground stations in the form of raw data. Experts analyse these data, interpret them and deliver a final product depending on the needs of the customer. The raw data would not be useful in practice without the interpretation and application of various filters that gives them commercial value. In fact, the Principles Relating to Remote Sensing of the Earth from Outer Space provide that information arrives on Earth as "primary data", and then through the use of sophisticated processes, it becomes "processed data" and finally, after interpretation, "analysed information". Thus, human factors play an important role in image interpretation.

Digital image processing involves the manipulation and interpretation of digital images from remotely sensed data using digital image-processing software techniques.[131] Digital image processing involves the following: (a) remote sensing data

[128] International Coalition for the Responsibility to Protect, Key development on the Responsibility to Protect at the United Nations from 2005–2017, http://responsibilitytoprotect.org/index.php/about-rtop/the-un-and-rtop, (accessed 23.09.2019).

[129] Gilmour, A., The Future of Human Rights: A View from the United Nations, pp. 247 in: Cambridge University Press, pp. 239–250, 2014, DOI: https://doi.org/10.1017/S0892679414000240, https://www.un.org/sg/sites/www.un.org.sg/files/atoms/files/Gilmour-FutureOfHumanRights.pdf, (accessed 23.09.2019).

[130] AAAS American Association for the Advancement of Science, Remote Sensing for Human Rights, https://www.aaas.org/programs/scientific-responsibility-human-rights-law/past-projects/remote-sensing-human-rights, (accessed 23.09.2019).

[131] United Nations Committee in the Peaceful Uses of Outer space, General Assembly A/AC.105/765/ 14 August 2001, Report on the United Nations/European Space Agency/Committee

import; (b) radiometric correction; (c) geometric correction; (d) image enhancement; (e) image classification; (f) map generation (for output of maps). In addition, for high accuracy geometric correction, Ground Control Points (GCP) are indispensable. However other methods are being proposed and developed to improve efficiency and to reduce the total amount of GCPs required for geometric correction.[132] There are different band frequencies of the electromagnetic spectrum that provide the interaction mechanism between the electromagnetic radiation and the material being imaged. Each image depicts an array of numbers indicating brightness levels (pixels). Image processing starts with radiometric and geometric correction to the image which then can be registered with a system of map coordinates so that other spatial data can be added. For better interpretation and analysis, an image is further enhanced and classified into various categories of object. Various enhancement techniques and classification tools can be provided.[133]

1.2.2.3 Certification of Space Data Processing

The challenges for the utilization of satellite remote sensing data are the potential abuse of privacy, human rights and public order. However, such data has an immense potential to facilitate human security and economic development. It can be reasonably stated that in certain conditions, EO data generated via remote sensing satellites provide objective data sources and a valuable source of information for decision makers. However, two questions arise with regard to the procedure for interpretation. Is all data from remote sensing objective? How can such data be admitted as a valuable source of information of its own in a court of justice?

Currently, experts can gain a remote sensing professional certificate to prove they are qualified to use the remote sensing professional software. Such training covers skills such as: satellite images mosaicking; convulsion filters; resolution merge; various enhancement techniques; creating contours; calculating slopes and aspects; shaded relief; 3D viewing for satellite images and data layers; creating flight paths through 3D satellite images; adding 3D models representing buildings, trees, aeroplanes, simulating floods and water movement; intervisibility analysis; adding fog, mist, and others to make realistic 3D effects on satellite images.[134] The objective of

on Space Research Workshop on Data Analysis and Image-Processing Techniques, Paragraph 24, pp. 6, http://www.unoosa.org/pdf/reports/ac105/AC105_765E.pdf, (accessed 22.09.2019).

[132] United Nations Committee in the Peaceful Uses of Outer space, General Assembly, A/AC.105/746, 15 January 2001, Report on the United Nations/ European Space Agency/Committee on Space research Workshop on Satellite Data Reduction and Analysis Techniques, para. 28, pp. 8, http://www.unoosa.org/pdf/reports/ac105/AC105_746E.pdf, (accessed 22.09.2019).

[133] Ibid. *supra* note General Assembly, A/AC.105/746, http://www.unoosa.org/pdf/reports/ac105/AC105_746E.pdf, para 4, pp. 10.

[134] Remote Sensing Professional Certificate https://remote-sensing-portal.com/certificates/rsp/, (accessed 23.09.2019).

obtaining such certification is for organization purposes, thus to identify and recognize those experts considered to meet the requirements to work with remote sensing data and maintain high standards of ethical conduct.[135]

Certification is only available and used to certify interested individuals in becoming experts in remote sensing analysis. However, the question that may be posed is whether the so-called training institutions are providing training following a standardized method. It could be reasonably argued that currently, the processing of remote sensing imagery is not done by following a standardized model. In practice, each of the so-called training institutions and experts that award and gain such accreditation works by different methods. The remote sensing community does not have a generally accepted and/or global approved standard.

The processed image cannot be certified for accuracy and authenticity because there is simply no such institution, whether at national or international level. This means that the processing of a satellite image and the final product is left to the analysis and advice of the expert, without any institution being able to certify that such work is accurate. In practice, the term “certified resellers” is being used to show the network of experts agreed by satellite operators to sell their images, but this does not replace the need for an organization that would certify the authenticity of remote sensing satellite images as final products. The lack of an independent organization at national or international level may be the reason why international courts and tribunals have chosen to work with in house experts or directly with the UN. Based on the ICC jurisprudence, satellite imagery processing is carried out by an in-house, not external, expert, in the person of Mr. Lars Bromley. Mr. Bromley is a satellite imagery expert and has been used to testify in different cases. An expert in digital evidence was used by the Office of the Prosecutor of the ICC for its Scientific Response Unit to collect and analyse space data imagery and this situation was recalled in literature as a means of enhancing the judicial institution’s independence as well as its reliability for the analysis and interpretation of satellite imagery. Also, it was well underlined that such a situation should be taken into consideration for the Post 2030-Agenda in terms of peace, justice and human rights. Thus, international courts could extensively contribute to the creation of international standards and methodologies for use, and integration of remote sensing satellite imagery as evidence.[136]

The practice of international courts regarding the use of satellite imagery will be analysed in a further chapter. However, it is worth underlining that because there are no specific rules regarding satellite imagery, its probatory value is analysed in comparison with the digitalized nature of proof. This is specific for satellite imagery and not for UAV and aviation pictures that do not need the same interpretation. In comparison with GNSS data whose receiver is immediately able to decode the signal received on his device without any processing, satellite imagery delivers only

---

[135] ASPRS The imaging and Geospatial Information Society, ASPRS Certification Program, https://www.asprs.org/Certification, (accessed 22.09.2019).

[136] Martin, A.S., Satellite Data as Evidences Before the Mechanism of International Courts, in: Annette Froehlich (Eds.) Post 2030-Agenda and the Role of Space: The UN 2030 Goals and Their Further Evolution Beyond 2030 for Sustainable Development, Springer International Publishing AG, part of Springer Nature 2018.

raw images that have to be processed. Until a national agency and/or an international organization exists to certify satellite imagery, international court judges need experts, trained and certified on the interpretation of these data. Such analysis is also important to avoid any error and verify if human rights have been violated or not. In this case, the work of the International Law Association Space Law Committee on the use of satellite data in courts is of particular relevance.

1.2.2.4 Legal Aspects of Data Collection

One of the most important legal aspects of data collection is privacy. The salient issues include what kind of data can actually be collected and distributed and who can have access to it. The matter is complex as the regulations may or may not be in place depending on who is collecting and what they are observing.[137] With different legal meanings and different levels of protection, "privacy" is a complex notion. It has been dealt with by national laws and by the EU as a whole. A reference to the right to privacy can be found in Article 12 UDHR.[138] This declaration has no binding legal force but is usually recognized as having customary value. This principle was repeated in Article 17 of the 1966 International Covenant on Civil and Political Rights. At the European level, the ECHR protects privacy in Article 8. The protection is against intrusions by public authority. Detailed legislation was developed at the national level. With respect to the EU, some regulations may be of particular importance for the right to privacy.

It was mentioned that in most countries, national legislation mostly protects against violations of privacy by public authorities, but no mention is made of private or commercial intrusions. These cases then need to be dealt with on a case-by-case basis, which does not offer an environment of legal certainty. Interestingly, cases such as privacy infringements by Google Street View have been dealt with differently in different European states. While Germany declared Google Street View as legal, Switzerland courts ruled that Google must take action to protect privacy, for example, by blurring out faces and licence plate numbers. In some countries, the term "privacy" does not even translate directly and is not protected as a right as such.[139]

Regarding remote sensing data, the issue of privacy is even more complex and less regulated. Not even public authorities' use of satellite technology is clearly regulated yet, let alone commercial or private use. An example was given in practice when a town in the USA used Google Earth to find swimming pools constructed without

---

[137] Schoenmaker, A., Community Remote Sensing Legal Issues, April 2011, pp. 4, https://swfound.org/media/62081/schoenmaker_paper_community_remote_sensing_legal_issues_final.pdf, (accessed 23.09.2019).

[138] Article 12 UDHR: "*No one shall be subjected to arbitrary interference with his privacy, family, home or correspondence, nor to attacks upon his honor and reputation. Everyone has the right to the protection of the law against such interference or attacks*".

[139] Schoenmaker, A., Community Remote Sensing Legal Issues, April 2011, pp. 4, https://swfound.org/media/62081/schoenmaker_paper_community_remote_sensing_legal_issues_final.pdf, (accessed 23.09.2019).

permits and to collect fines. After much negative press, the local council decided to stop this practice through satellite images, but there was no law in place to force it to do so. Another widely reported case concerns the ability of police or other public authorities to track the location of a person through a cell-phone, with the help of cellular phone service providers. Whether this should be allowed or not is much debated, as it can have many applications, both positive (finding a missing person in need of help or locating a criminal) and negative (helping an abusive husband or parent find a person who is trying to escape).[140] It can be reasonably stated that privacy aspects would probably not represent a legal issue for most of the data from remote sensing that is being used publicly. The laws, regulations and public feelings about it are so complex and diversified, that this uncertain situation might actually inhibit the use or collection of certain data.

1.2.2.5 Data Access and Usage

It is generally acknowledged that access to existing data is relatively effortlessness, using the Internet, online data bases or even free access to programs such as Google Earth. A classification of data access was made in practice based on the source (free access, commercially available); on the customer (government, commercial, education institutes, private etc.); and on the purpose (disaster mitigation, climate change, business development, education etc.). Re-using the data for another purpose imposes a set of rules, even for data that is freely available.[141] Data should be forbidden to be modified and added upon by anyone. The risk of inaccuracy of volunteer-based projects such as Wikipedia or other participation on "open" projects should be not allowed to be used in courts. In such cases there should be investigation as to who takes responsibility, if any, for the conformity of the processed data.

Copyright applies only to processed data. Raw data is not an original creation and thus it is not the appropriate way to protect remote sensing data. Instead, database protection could apply to raw data or ownership rights. Directive 96/9/EC of the European Parliament and of the Council on the legal protection of Databases was adopted in March 1996, and provides copyright protection for the intellectual creation involved in the selection and arrangement of materials; and sui generis protection for an investment (financial and in terms of human resources, effort and energy) in the obtaining, verification or presentation of the contents of a database.[142]

---

[140] Ibid. *supra* note Schoenmaker, A., Community Remote Sensing Legal Issues, April 2011, pp. 4, https://swfound.org/media/62081/schoenmaker_paper_community_remote_sensing_legal_issues_final.pdf, (accessed 23.09.2019).

[141] Schoenmaker, A., Community Remote Sensing Legal Issues, April 2011, pp. 6, https://swfound.org/media/62081/schoenmaker_paper_community_remote_sensing_legal_issues_final.pdf, (accessed 23.09.2019).

[142] Ibid. *supra* note Schoenmaker, A., Community Remote Sensing Legal Issues, April 2011, pp. 6, https://swfound.org/media/62081/schoenmaker_paper_community_remote_sensing_legal_issues_final.pdf, (accessed 23.09.2019).

For using remote sensing data, policies are usually in place, which in most cases require a license, including restrictions such as: the data shall be used only for the particular purpose for which the license is granted; it shall not be used for any purpose that would be against the law; it shall not be altered; it shall not be further distributed.[143] Alteration of the processed data from remote sensing could carry liability and the question is, who will decide about the authenticity of the processed data in the absence of a national agency or international institution with such powers? The problem of the authenticity of processed data is different from the error data that could be provided in exceptional cases by the GNSS providers, because the first case is about an illegal human intervention over the processed data while the second case is about a system error. GPS is a free service but has never been guaranteed by the U.S. Government and the same issue could apply to GALILEO, which is the European GNSS.

1.2.2.6 New Technologies for Surveillance of Criminal Offences

Surveillance technology has evolved, and public order authorities wish to make use of it to become more effective in combating crimes and violations. The scope of this subchapter is to prove how useful the utilization of new technology is, and to describe the effort to implement such technology at national and local levels, and the challenges or limitations of such technology.

The police were engaged in other high-priority calls when they received a report of two people fighting. "Drones are responding to 911 calls in this California city" was the title of a recent article in CNN News about police using drones (UAV) to respond to emergencies. A police drone flew to the site and broadcasted a live video feed to an officer's smartphone. The drone footage was useful to the police officers, at arrival they already had initial situation awareness minutes before and could make the arrest based on the images. This is just an example of how Chula Vista Police Department reacted to criminal offences using drones as part of a Federal Aviation Administration Program.[144]

In 2017, the U.S. Transportation Department established the UAS Integration Pilot Program that brings UAS operators and manufacturers together with state and local governments to accelerate safe drone integration. Under its Integration Pilot Program, ten awardees were selected, each with a particular project that covers a broad type of operations in urban and rural areas.[145] This program will support the Federal Aviation Administration (FAA) working with local, state and tribal governments as well as

---

[143]Ibid. *supra* note Schoenmaker, A., Community Remote Sensing Legal Issues, April 2011, pp. 6, https://swfound.org/media/62081/schoenmaker_paper_community_remote_sensing_legal_issues_final.pdf, (accessed 23.09.2019).

[144]Chen, N., Drones are responding to 911 calls in this California city, 12 April 2019, https://edition.cnn.com/2019/04/12/us/california-drones-emergency-response/index.html, (accessed 23.09.2019).

[145]Federal Aviation Administration, Integration Pilot Program Lead Participants, 23 April 2019, https://www.faa.gov/uas/programs_partnerships/integration_pilot_program/lead_participants/, (accessed 23.09.2019).

business to see how the integration of drone operations could successfully support police operations, inter alia.[146] The data gathered from this pilot project will support a new regulatory framework to safely integrate drones in the national airspace. The UAS Integration Pilot Program (IPP) was designed to helping the US Department of Transportation (USDOT) and FAA craft new rules.[147] There are four main goals in the UAS IPP program:

(1) Identifying ways to balance local and national interests related to drone integration;
(2) Improving communications with local, state and tribal jurisdictions;
(3) Addressing security and privacy risks;
(4) Accelerating the approval of operations that currently require special authorizations.

Furthermore, the evaluation of the activities includes night operations, flights over people and beyond the pilot's line of sight, package delivery, detect-and-avoid technologies and the reliability and security of data links between the pilot and aircraft.[148] Such operations are an extension of the rules and limits imposed on recreational flyers or other usual UAV operations by the FAA published in the "Public Safety and Law Enforcement Toolkit" designed to "assist law enforcement and public safety entities in operating and handling situations involving drones or UAS".[149] Intel is supporting the testing of this technology stated that, "States and localities will continue to look to drones for safer, more efficient infrastructure inspection, search and rescue and disaster relief".[150]

This activity is apposite as it addresses security and privacy risks. The Chula Vista Police Department (CVPD) is at the "forefront of utilizing drones to enhance the science of policing" and the results from the pilot project are expected by other agencies across the USA to replicate the use of drones for emergency response and support.[151] The success of the first period was mentioned by the Chula Vista Chief of Police who said that the drone program "is already having a significant impact

[146]Zazulia, N., DOT Selects 10 Participants for Nationwide Drone Integration Pilot Program, 11 May 2018, https://www.aviationtoday.com/2018/05/11/dot-selects-10-participants-nation-wide-drone-integration-pilot-program/, (accessed 23.09.2019).

[147]Federal Aviation Administration, UAS Integration Pilot Program, 7 November 2019, https://www.faa.gov/uas/programs_partnerships/integration_pilot_program/, (accessed 23.09.2019).

[148]Ibid. *supra* note, Federal Aviation Administration, UAS Integration Pilot Program, 7 November 2019, https://www.faa.gov/uas/programs_partnerships/integration_pilot_program/, (accessed 23.09.2019).

[149]Federal Aviation Administration, Public Safety and Law Enforcement Toolkit, 4 September 2019, https://www.faa.gov/uas/public_safety_gov/public_safety_toolkit/, (accessed 23.09.2019).

[150]Ibid. *supra* note Zazulia, N., DOT Selects 10 Participants for Nationwide Drone Integration Pilot Program, 11 May 2018, https://www.aviationtoday.com/2018/05/11/dot-selects-10-participants-nation-wide-drone-integration-pilot-program/, (accessed 23.09.2019).

[151]Zazulia, N., IPP Drones Aid in Arrests for California Police Department, 12 February 2019, https://www.rotorandwing.com/2019/02/12/ipp-drones-aid-arrests-california-police-department/, (accessed 23.09.2019).

on operations and resource management. Real time aerial visibility is critical when informing decisions, and, in an emergency situation, is vital to the safety of our officers and citizens". The CVPD describes the mission of the UAS program "to provide airborne support to police operations in a safe, responsible, and transparent manner to preserve the peace, reduce response times." In particular, the use of UAS drones by the police department is "an efficient and effective way of providing law enforcement critical information to respond to Calls For Service, emergency situations or to conduct criminal investigations", such as providing an overhead view of an area or incident for ground personnel, safely clearing the interior of buildings, providing detailed documentation of crime and accident scenes, and searching for lost or missing persons.

A special focus was to address concern about public trust and the public's right to privacy during the operation of CVPD UAS systems. It was clearly underlined that to respect civil liberties and personal privacy, the UAS policy "specifically prohibits the use of UAS for general surveillance or general patrol operations". The activity consists in deploying drones from the rooftop of the Police Department Headquarters after 911 calls. The CVPD UAS is used a first responder to emergency incidents such as crimes in progress, fires, traffic accidents and reports of dangerous subjects. The support that will be given to police officers is to present a faster perspective:

> This Drones as a First Responder (DFR) System is transformational by providing first responders with something they have never had before, a faster perspective of the situation since the drones are deploy to the incident and arrive well before ground units. The on-board camera streams HD video back to the department's real-time crime centre where a teleoperator, who is a trained critical incident manager, not only controls the drone remotely, but communicates with the units in the field giving them information and tactical intelligence about what they are responding to. The system also streams the video feed to the cell phones of the first responders and supervisors on the ground so they can see exactly what the drone is seeing. The drones have responded to hundreds of calls for service giving Chula Vista Police Officers information that no other public safety agency has ever had before; aerial intelligence about what exactly they are responding to before they arrive. Imagine the value of knowing that the truck leaving the scene of a robbery report is red and heading northbound, or that the report of a man with a gun is actually a 16-year-old with a BB gun, or the accident on the freeway involves a tanker truck with placards indicating a chemical hazard. These are things traditional manned air support has not been able to provide, as they are more commonly called in by ground units once they have arrived. Widespread deployment of DFR Systems will transform the way public safety agencies serve their communities, increasing safety, efficiency, and accountability.[152]

France is another example where airspace imagery is used for crime investigation. As part of the "new high-tech weapons against rioters" the riot police are equipped with drones/UAVs which can provide photos and video footage of the crowds.[153] Such equipment helps to identify solutions and to prosecute violent protesters and

---

[152] Chula Vista Police, UAS Drone Program, https://www.chulavistaca.gov/departments/police-department/programs/uas-drone-program, (accessed 23.09.2019).

[153] The Local Newspaper, Drones and UV spray: French police to get new high-tech weapons against the rioters, 19 March 2019, https://www.thelocal.fr/20190319/drones-and-uv-spray-frances-new-high-tech-weapons-against-rioters, (accessed 09.06.2019).

looters. The UAVs will eventually become autonomous but the outcome from current investigations is that the UAVs are used locally, they have a limited range and in-flight duration, and need to be deployed and controlled by the police. They are helpful and could be used by all police forces around the globe to detect and prosecute offenders. However, the UAVs only represent a preliminary step in the large-scale utilization of remote sensing from satellites, particularly the technology of small satellites, such as those part of large constellations of satellites.

In general, public perception is that images from space mainly come from spy satellites. However, it should be noted that access to spy satellites is only allowed to those who poses such technology and are authorized for analysis as official agents. The specialized institutions eventually use such information as part of their reconnaissance activity and disseminate information for internal use only and not to be utilized as evidence in a public court. It is important for the current research to understand that analysis of satellite imagery exists, it is used and is analysed by experts. A particular case of satellites fighting crime was when satellite imagery was used to assist British police search for the body in a murder case. Aerial images were not adequate for the investigators to analyse wide areas and so they used satellite imagery provided by Satellite Applications Catapult in the Rania Alayed case. The analyses from satellite imagery enabled the police "to identify anomalies on the ground, which we could search straight away".

The representative of Greater Manchester Police positively referred to satellite technology as the "eye in the sky" and underlined that this technology saved them months of fruitless work: "We see this technology as taking us into the next generation of crime investigation. (…) This is moving forward".[154] Referring to the improvement of remote sensing satellites resolution in 2014, Dr. Richard Hilton, representative of SA Catapult, revealed that for the Rania Alayed case, they used a 30 cm resolution: "This year we have 30 cm resolution, which dramatically changes the potential for detecting things, or monitoring a site of interest". The Australian Authorities used remote sensing technology in criminal investigation even before 2014.[155] But technology advances are now enabling far more accurate and reliable imagery that could revolutionize 21st-century policing, transforming law enforcement capacity through highly detailed surveillance.

It is reasonable to expect that small satellites will become the technology to provide cost effective solutions for imaging and video. In line with the prediction from Ray Purdy, expert on satellites and the law at University College London, it could be cost effective for police forces to buy their own satellites.[156] Purdy has been able to

---

[154]Monks, K., Spy satellites fighting crime from space, 12 August 2014, https://edition.cnn.com/2014/08/11/tech/innovation/spy-satellites-fighting-crime-from-space/index.html, (accessed 9.06.2019).

[155]Ibid. *supra* note Monks, K., Spy satellites fighting crime from space, 12 August 2014, https://edition.cnn.com/2014/08/11/tech/innovation/spy-satellites-fighting-crime-from-space/index.html, (accessed 9.06.2019).

[156]Purdy, R., Using Earth Observation Technology for Better Regulatory Compliance and Enforcement of Environmental Laws, In: J. of Environmental Law, 22(1), February 2009, DOI: https://doi.org/10.1093/jel/eqp027, (accessed 26.09.2019).

monitor criminal activity, such as large-scale illegal waste disposal, through satellite surveillance, which would have allowed police to cheaply and easily strengthen existing cases.[157]

1.2.2.7 Spatial Data Infrastructure and Use

Public access to remote sensing data collected by satellites was until recently very limited. Human rights advocacy groups are increasingly relying on satellite imagery to examine events in violation of human rights and many examples will be given in the section that analyses cases of human rights violations. The use of satellite-collected images in human rights investigations reflects a number of opportunities but also challenges and risks. It has been argued that if satellites are to be used effectively to collect evidence, then greater attention to their capacity for ensuring accountability is required from institutions and justice.[158]

The term "satellite imagery" extends to a number of digital images that are usually referred to as data. For this reason, the question arises as to whether satellite images are really pictures. It can be reasonably ascertained that satellite images constitute data, which is, for the matter discussed here, substantially different. Advantages of satellite imagery include its capability to cover much greater areas, have the potential to offer real-time streaming and global coverage, with better prices. Because all information is digital, it can be easily transmitted.[159] By contrast, aerial photography includes the production of photographic images from the sky. Balloons, planes, helicopters, drones or even submarines are all capable of remote sensing and such technology has been used for investigations. Similar technology could be successfully utilized also in the context of human rights.[160] Offering similar accuracy for the images, aerial photography is easier to understand and eliminates the need for special analysis and interpretation. However, it covers limited areas and is available only on demand. Indeed, if a picture is altered, it will not be hard to prove it when referring to UAVs, planes, balloons etc. However, from the literature on the matter the same cannot be said with regard to satellite images. In fact, they are a gathering of data that can be altered, without it being possible to prove the alteration once the

---

157 Ibid. *supra* note Monks, K., Spy satellites fighting crime from space, 12 August 2014, https://edition.cnn.com/2014/08/11/tech/innovation/spy-satellites-fighting-crime-from-space/index.html, (accessed 23.09.2019).

158 Notley, T., Webb-Gannon, C., Visual Evidence from Above: Assessing the Value of Earth Observation Satellites for Supporting Human Rights, 2016, http://twentyseven.fibreculturejournal.org/2016/03/21/fcj-201-visual-evidence-from-above-assessing-the-value-of-earth-observation-satellites-for-supporting-human-rights/, (accessed 5.06.2019).

159 Kotz, S., What is the Difference Between Satellite Imagery & Aerial Photography, 13 March 2018, https://sciencing.com/difference-satellite-imagery-aerial-photography-8621214.html, (accessed 23.09.2019).

160 Ibid. *supra* note Notley, T., Webb-Gannon, C., Visual Evidence from Above: Assessing the Value of Earth Observation Satellites for Supporting Human Rights, 2016.

final image is produced. Therefore, in terms of its possible use as legal evidence, satellite imagery poses a different challenge when compared to regular pictures.[161]

The number of particularities for spatial data continue with regards to the matter of sovereignty. Satellites are outside the sovereign borders of each country, thus no authorization is needed from the sensed states, and satellites can be used to capture information from the most remote areas of the Earth. Satellite imagery is part of digital technologies. In fact, its particularity is to gather evidence via remote sensing, based on satellite technology. Different sensor types can be applied to transform the raw data in images, and these include optical, radar and infrared sensors.[162]

1.2.2.8 Technical Capabilities

What can be detected, recognized and identified through the use of remote sensing depends on the technical capabilities of Remote Sensing Satellites in delivering satellite imagery.[163] The three major characteristics of remote sensing data are: (i) spatial information; (ii) spectral information; (iii) temporal information. Spatial resolution is between 10 m and 1 m ground resolution. Spectral resolution involves different filters being applied on the images. Temporal resolution allows comparison and analysis between different dates and multi-year observations that result in a time series. The conclusion is that a large amount of information can be collected from space.

Satellite systems for remote sensing are of two main categories: optical and radar. Their spectral and spatial resolution depends on the specific sensor, with a range of pixel sizes from tens of meters to less than half a meter. It was mentioned that optical satellites can allow a three-dimensional representation of the ground surface due to multiple images being acquired from different angles to produce a Digital Elevation Model (DEM). Referring to optical satellites, their operation is typically in the visible and near-infrared part of the spectrum but can be extended to multispectral capabilities. However, its capabilities of acquiring data in a number of wavelength bands not visible to the naked eye is mainly used for information about vegetation and lithological characteristics including mineralogy. The multispectral bands are typically acquired at lower resolutions than the visible wavelengths.

Whereas optical satellites rely on reflected or emitted electromagnetic radiation initially sourced from the sun, radar sensors on satellites are active and send pulses of microwave energy to the earth's surface and record the return signal. Individual radar sensors, such as TerraSAR-X, operate at a particular wavelength and offer

[161]Nunez, M.A.C., Admissibility of remote sensing evidence before international and regional tribunals: Innovations in Human Rights Monitoring Working Paper, August 2012, pp. 3, https://www.amnestyusa.org/pdfs/RemoteSensingAsEvidencePaper.pdf, (accessed 23.09.2019).

[162]Ibid. *supra* note Nunez, M.A.C., Admissibility of remote sensing evidence before international and regional tribunals: Innovations in Human Rights Monitoring Working Paper, August 2012, See also Johnson 2013.

[163]Mostert, S., Witt L., "Technical Capabilities of Remote Sensing Satellites: The Potential for Human Scale Development or Abuse" in Proceedings of the International Institute of Space Law 2014.

high resolutions down to 0.25 m. These sensors are independent of cloud coverage, sensitive to identifying surface characteristics including texture and geomorphology, whilst also allowing DEM generation.[164]

1.2.2.9 Admissibility of Remote Sensing Evidence Before International and Regional Tribunals

A Report on satellite imagery as evidence in legal and administrative proceedings, prepared for ESA in 2012 by the London Institute of Space Policy and Law, highlights a number of aspects that were summarized as follows: (i) the gap in the legal, technical and administrative communities in understanding the use of Earth Observation (EO) information; (ii) the need to engage and inform the legal, technical and administrative communities; (iii) the existing standards and procedures do not address the use of space-derived EO information before courts and administrative tribunals.[165] The practicality of such research comes from setting the stage and applying evidential rules in selected jurisdictions that enable analysing existing standards and procedures and making proposals for the better use of satellite imagery as evidence. This research focused on British judicial law, explaining this selection in relation to the "highly demanding evidential rules" required by the common law system that makes the analysis relevant to jurisdictions with the strictest evidence requirements. Choosing the law of evidence included analysis of the rules of admissibility, rules that qualify the weight to be given to specific evidence, and the standard of proof that governs the degree of certainty required for a decision. Also, the choice of reference law has taken into consideration that common law and civil law jurisdictions differ as to the admission of evidence.[166]

164 Airbus, Oil, Gas, Mining and Energy. Earth Observation for Exploration, Development and Production, http://www.airbusds-ogm.com/news/supporting-exploration-with-satellite-imaging-technology/, (accessed 23.09.2019).

165 Document ESA-ISPL/EO 76/final, London Institute of Space Policy and Law, (2012) Evidence from Space. Study for the European Space Agency on Use of Space-Derived Earth Observation Information as Evidence in Judicial and Administrative Proceedings, pp. 9, https://www.space-institute.org/app/uploads/1342722048_Evidence_from_Space_25_June_2012_-_No_Cover_zip.pdf, (accessed 20 September 2019).

166 *Common law and civil law jurisdictions differ as to the admission of evidence. Broadly, civil law countries have a more inclusive approach, while common law jurisdictions may take a more exclusionary approach. This is to some extent due to the role of the judge in the respective systems. In the inquisitorial model associated with civil law systems, judges have a major role in uncovering the facts. This tendency is exemplified by the role of the juge d'instruction in France, who examines witnesses before pronouncing whether or not the evidential basis exists for a case to go to trial. The judge has wide discretion to admit or reject evidence, and concentrates more on the relevance of evidence than on rules of admissibility. By contrast, the common law applies stricter rules on admissibility. In the adversarial model associated with common law systems, advocates for the parties take the lead in presenting and testing evidence. The role of the judge is more limited than in civil law jurisdictions. Rules of evidence have evolved within common law systems to ensure that only relevant and probative evidence is admitted.*

In assessing the probatory value of satellite data as evidence, their interpretation and the availability of EO imagery, and in the absence of any universal legal framework for the use of such information, it was underlined that reference should be made to national laws and rules of evidence governing reliability requirements in respect of electronic records. Such rules refer to the suitability of the system for the creation of electronic records and the possibility of verification by an accepted certification process, or by evidence of calibration, which by analogy would be applicable to EO information.[167] The study identified instruments that specifically mention or indicate the use of EO information as evidence.[168] However, such rules were described as falling outside of the analysis because they "either do not rely primarily on civil courts or administrative tribunals for enforcement, or the applicable arrangements do not involve courts or administrative systems for other reasons".[169]

Special consideration must be given to the burden of proof in analysing data from remote sensing satellites to prove violations of human rights. A principle of criminal law is that a person should be presumed innocent unless proven guilty. Such a principle was explained as referring to the State preference "to run the risk of acquitting people who may in fact have committed the offence than convicting the people who are in fact innocent".[170] The presumption of innocence in criminal law is reinforced by Article 6 (2) ECHR "Right to a fair trial" which declares "Everyone shall be presumed innocent until proved guilty according to the law".

Analysis of this principle has concluded that the ECHR has not interpreted the presumption "as strictly as it might".[171] It was shown in case of Salabiaku v. France[172] and in the case of Philips v. UK[173] that "the presumption of guilt or the placing of the burden of proof on the defendant, in other words a reverse burden of proof,

---

[167] Ibid. *supra* note Document ESA-ISPL/EO 76/final, London Institute of Space Policy and Law, (2012) Evidence from Space. Study for the European Space Agency on Use of Space-Derived Earth Observation Information as Evidence in Judicial and Administrative Proceedings, pp. 18.

[168] Ibid:

(1) *Spatial information infrastructures;*
(2) *Arms control and verification;*
(3) *Nuclear proliferation;*
(4) *Natural disasters and humanitarian relief;*
(5) *Restrictions on the resolution of imagery;*
(6) *Marine pollution;*
(7) *Claims for agricultural subsidies; and*
(8) *Cadastral mapping and support of agricultural development.*

[169] Ibid. *supra* note Document ESA-ISPL/EO 76/final, London Institute of Space Policy and Law, (2012) Evidence from Space. Study for the European Space Agency on Use of Space-Derived Earth Observation Information as Evidence in Judicial and Administrative Proceedings, pp. 21.

[170] Ibid. *supra* note Herring, J., Criminal Law. Text, Cases and Materials, in: Chap. 1 *An Introduction to Criminal Law,* pp. 46 Reference to Lord Sankey LC in Woolmington v DPPI [1935] AC 462, 481 (HL), Oxford University Press, 5th Edition, 18 June, 2012.

[171] Ibid. *supra* note Herring, J., Criminal Law. Text, Cases and Materials, in: Chap. 1 *An Introduction to Criminal Law*, pp. 47, Oxford University Press, 5th Edition, 18 June, 2012.

[172] (1988) 13 EHRR 379.

[173] [2001] Crim LR 217.

does not contravene Article 6 ECHR if it is confined within "reasonable limits". The reverse burden of proof requires analysis of what is "a persuasive burden of proof" (the party must prove the matter at issue) and what is meant by an "evidential burden of proof" (means that the party must introduce sufficient evidence to establish "the reasonable possibility" that a particular issue is true). Thus, the Court has the obligation to consider all the aspects of the case and the particularities of gathering evidence. As a consequence, the Court will need to consider the relative difficulty of proving the relevant matter and the seriousness of the threat at which the criminal offence is directed. The conclusion was that a persuasive burden would be justifiable only if it involves an issue that is very easy for the defendant to establish,[174] (e.g. to prove that he or she has a license to perform an activity[175]) or the burden is necessary for the protection of the public. In the light of the Humans Rights Act, Section 3, which requires "So far as it is possible to do so, primary legislation and subordinate legislation must be read and given effect in a way which is comparable with the Convention rights" any statutory burden of proof on the defendant would be interpreted to be evidential if at all possible.

The scope of application of satellite data in administration and justice becomes necessary for effective, accountable and transparent decision making.[176] Space technology and policy have been used to promote growth, stability, security and peace on a global level, thus space activities benefit social and economic activities on Earth and such benefits should be transferred also in justice. Thus, it is a priority to "enlarge the scope of application of satellite data" in administration and justice and provide the necessary regulatory support for such innovative mechanisms.[177] Also, if satellite imagery was used extensively as evidence in courts and if the understanding of remote sensing was more widespread, it could become a mechanism to support the Post-2030 Agenda SDG in relation to space society.[178] The importance of the use of satellite data as evidence before international courts was debated in some papers. It analysed for example the applicable law and/or presented international case law. Such research also supports the view that, for the purposes of the Post 2030-Agenda, it can be envisaged that satellite imagery will become more widespread and will

---

[174] Ibid. *supra* note Herring, J., Criminal Law. Text, Cases and Materials, in: Chap. 1 *An Introduction to Criminal Law*, pp. 47, Oxford University Press, 5th Edition, 18 June, 2012.

[175] Compare DPP v Barker [2006] Crim LR 140 and R v Makuwa [2006] Crim LR 911.

[176] Froehlich, A. (Eds.) Post 2030-Agenda and the Role of Space: The UN 2030 Goals and Their Further Evolution Beyond 2030 for Sustainable Development, Springer International Publishing AG, part of Springer Nature 2018.

[177] Ibid. *supra* note Froehlich, A. (Eds.) Post 2030-Agenda and the Role of Space: The UN 2030 Goals and Their Further Evolution Beyond 2030 for Sustainable Development.

[178] Committee on the Peaceful Uses of Outer Space A/AC.105/C.2/2019/CRP.10, Zero draft, The "Space2030 Agenda: Space as a driver of sustainable development", 29 March 2019, http://www.unoosa.org/res/oosadoc/data/documents/2019/aac_105c_22019crp/aac_105c_22019crp_10_0_html/AC105_C2_2019_CRP10E.pdf, (accessed 20.09.2019).

be significant both for traditional issues such as environmental protection or border delimitation, but also for human rights protection.[179] With the increasing and streamlining use of data and information derived from space, satellite imagery has the potential to provide access to evidence from remote and hardly accessible areas and help in prosecuting human rights violations.[180]

In relation to SDG 16 (Peace, Justice and Strong Institutions) and SDG 17 (Partnerships for the Goals), satellite imagery would be useful to support and build "just, equitable and more inclusive institutions". Therefore, satellite imagery could represent an important technical element in gathering evidence of international crimes, especially in areas where it may be difficult to gather evidence.[181] The use of satellite imagery in trials, including human rights violations, could be reasonably regarded as the next step in the progress of international justice.

The collection and assessment of evidence is strictly determined in the judicial procedures applicable to international courts or tribunals. In this sense, the Report prepared for ESA in 2012 by the London Institute of Space Policy and Law mentions that the use of satellite data is a part of documentary evidence, and more particularly, it belongs to the category of digital or electronic evidence that can take the form of EO data or from the GNSS.[182] The acceptance and probatory value of satellite imagery depends on the regulations of courts. The 1986 Principles Relating to Remote Sensing of the Earth from Outer Space, adopted by UNGA, do not mention the possibility of using satellite imagery during lawsuits and do not provide any specific rule on this matter. In addition, neither space law treaties nor resolutions of principles provide any specific rule for using the GNSS as evidence. Therefore, in the absence of specific references to the use of satellite data during a legal proceeding, courts may find themselves legally constrained in giving probative value and making the most of satellite imagery. The necessary step is not only to highlight this legal gap but also to address it.

It is reasonable to state that using satellite data during trials is becoming unexceptional and highly realistic. There are many examples of this trend in practice including the cases of the ICJ, of the International Tribunal for the Law of the Sea, and of the ICC, which will be analysed further.[183] It also needs to be emphasized

---

[179] Ibid. *supra* note Martin, A.S., Satellite Data as Evidences Before the Mechanism of International Courts, in: Annette Froehlich (Eds.), Post 2030-Agenda and the Role of Space: The UN 2030 Goals and Their Further Evolution Beyond 2030 for Sustainable Development, Springer International Publishing AG, part of Springer Nature 2018.

[180] Ibid. *supra* note Martin, A.S., Satellite Data as Evidences Before the Mechanism of International Courts, pp. 98.

[181] Ibid. *supra* note Martin, A.S., Satellite Data as Evidences Before the Mechanism of International Courts, pp. 98.

[182] Ibid. *supra* note Document ESA-ISPL/EO 76/final, London Institute of Space Policy and Law, (2012) Evidence from Space. Study for the European Space Agency on Use of Space-Derived Earth Observation Information as Evidence in Judicial and Administrative Proceedings, pp. 28. See also: Martin, A.S., Satellite Data as Evidences Before the Mechanism of International Courts, pp. 98.

[183] Ibid. *supra* note Martin, A.S., Satellite Data as Evidences Before the Mechanism of International Courts, pp. 98.

that in fact, satellite imagery could be considered as a unique tool necessary to solve certain cases such as boundary disputes or environmental issues and also "to address the widespread and systematic character of such acts".[184] The protection of human rights is an important objective to which satellite imagery could contribute in detecting human rights violations. Overall, the issue of the admissibility of satellite data and its probative value could be made on a case-by-case basis, according to the principle of free assessment of evidence. Harmonization of the legal framework relating to satellite imagery as evidence before international courts could be regarded as necessary in respect of the advancements in remote sensing technologies that are, and will increasingly become, even more operated not only by states, but also by private entities.[185]

When referring to criminal law and human rights, the form of interpretation of the offence according to the Human Rights Convention must comply with the criminal law and procedures for evidence. The opposite is also true. For example, the UK Human Rights Act of 1998, Section 3 requires judges to interpret legislation in a way that complies with the ECHR, thus: "So far as it is possible to do so, primary legislation and subordinate legislation must be read and given effect in a way which is compatible with the Convention rights". The meaning of the phrase "so far as it is possible" refers to the possibility of the courts interpreting evidence in a way that is compliant with the Convention rights.[186]

---

[184] Ibid. *supra* note Martin, A.S., Satellite Data as Evidences Before the Mechanism of International Courts, pp. 98.

[185] Ibid. *supra* note Martin, A.S., Satellite Data as Evidences Before the Mechanism of International Courts, pp. 98.

[186] Herring, J., Criminal Law. Text, Cases and Materials, in: Chap. 1 *An Introduction to Criminal Law*, pp. 39, Oxford University Press, 5th Edition, 18 June 2012.

Chapter 2
Practical Use of Satellite Data in Support of Human Rights

Abstract This chapter debates the usefulness of satellite imagery from a practical perspective. Cases presented analyse the important role of satellite imagery in updating and extracting land related information, such as a historical record of the areas that are subject to changes over time and their usefulness for territorial delimitation. In analysing these cases, the work of non-governmental organizations for human rights will be highlighted including the use of satellite imagery in their investigations. Examples are provided in regards to the need to prove violations of human rights with the help of satellite imagery during wartime. Another aspect is how real time deforestation detection through satellite images could be among the game changing solutions to save the Amazon Rainforest and to defending indigenous communities. Further questions turn around why satellite imagery is indispensable for forecasting the weather and necessary for waste management, and why it is effective for oil spill surveillance. Finally, as per the treaty compliance requirements, the need for treaties to specifically provide for the use of satellite data as a tool to monitor compliance will be discussed.

2.1 Historical Consideration of Satellite Data

The use of satellite imagery can play an important role in updating and extracting land related information, it can provide valuable evidence such as a historical record of the areas that are subject to changes over time and prevent conflicts over boundaries that may also involve violations to human rights for cross-border movements.[1] Harsh conditions are encountered when doing traditional cadastral survey in remote areas, especially in mountainous areas.[2] Advances in space-based remote sensing have revolutionized the field of cadastral surveying and mapping because satellite imagery

[1] Wolfinbarger, S., Drake, J., Ashcroft, E., Monitoring Border Conflicts with Satellite Imagery: A Handbook for Practitioners, AAAS, pp. 4, 2015, https://www.aaas.org/sites/default/files/reports/Handbook.pdf, (accessed 28.09.2019).

[2] Ali, Z., Ahmed, S., Extracting Parcel Boundaries from Satellite Imagery for a Land Information System, Conference Paper: 6th International Conference on Recent Advances in Space

A. Froehlich and C. M. Tăiatu, *Space in Support of Human Rights*,
Studies in Space Policy 23, https://doi.org/10.1007/978-3-030-35426-8_2

can provide data for mapping applications and can be used as an alternative or a support to traditional land surveying.[3]

The delimitation process is a politically sensitive issue as it often challenges the sovereignty of states over strategic territories. The OSCE has referred to international boundary making and the lack of border demarcation or delimitation by stating that it constitutes an obstacle to border security and management but also a threat to the territorial integrity of states. Internationally recognized borders are necessary to facilitate the legal movement of persons and goods. It is important to support the peaceful settlement of border disputes.[4] Many new international borders have been created since 1989, such as those resulting from the collapse of the former Soviet Union and Yugoslav Republics but also in Africa and the complex hydrological relationships across Central Asia boundaries in the Mau Darya and Syr Darya river basins.[5] Best practices for boundary delimitation and demarcation processes can be provided by the UN but also the OSCE who both aim to be a neutral and impartial forum for discussions and to support the sovereign decision of states in reaching peaceful settlements.

International assistance in land and sea boundary delimitation/demarcation is necessary especially where there is a lack of expertise. Discussions at the OSCE have underlined these issues by stating that:

> (…) boundary-making efforts require a wide range of expertise, which many States around the world do not have access to. This situation can delay boundary-making efforts which can then increase the risk of possible disputes. The UN can provide impartial expertise that supports States in all phases of boundary making, including preparation, delimitation and demarcation. In the preparation stage, the UN Cartographic Section can assist in setting-up institutions, in facilitating dialogue between the parties and in the collection of cartographic material. In the delimitation phase, the UN Cartographic Section may be able to serve as a facilitator within a boundary commission and as the commission's documentary secretariat.[6]

The UN Cartographic Section provided technical assistance in international boundary issues in cases such as the Iraq-Kuwait boundary, the determination of the Line of Withdrawal (Blue Line) between Israel and Lebanon, and the Green Line in Cyprus. The UN also supported the work of the Eritrea-Ethiopia Boundary Commission

---

Technologies (RAST 2013), June 2013, pp. 79, https://www.researchgate.net/publication/239237780_Extracting_Parcel_Boundaries_from_Satellite_Imagery_for_a_Land_Information_System, (accessed 20.07.2019).

[3]Ibid. *supra* note. Ali, Z., Ahmed, S., Extracting Parcel Boundaries from Satellite Imagery for a Land Information System, Conference Paper: 6th International Conference on Recent Advances in Space Technologies (RAST 2013), June 2013, pp. 79.

[4]OSCE Organization for Security and Co-operation in Europe, Applied Issues in International Land Boundary Delimitation/Demarcation Practices, pp. 5, 2011, https://www.osce.org/secretariat/85263?download=true, (accessed 21 July 2019).

[5]Foucer, M., Institute of Higher National Defence Studies, Paris, France in: OSCE, Applied Issues in International Land Boundary Delimitation/Demarcation Practices, pp. 7, 2011, https://www.osce.org/secretariat/85263?download=true, (accessed 21.07.2019).

[6]EOM K.S., Cartographic Section, Department of Field Support, United Nations, New York, in: OSCE, Applied Issues in International Land Boundary Delimitation/Demarcation Practices, pp. 22, 2011, https://www.osce.org/secretariat/85263?download=true, (accessed 21.07.2019).

and the work to facilitate demarcation of the Cameroon-Nigeria boundary and the North-South Sudan Border Committee with the African Union.[7] Satellite data can support the procedures for base mapping which is essential for the analysis of the documentary material for decision making and implementation. Technical experts, advisers and decision makers should in all the situations make use of satellite data in order to have all the data available for the best possible geographical knowledge of the border.

A state company from France with expertise in map production and land assessment including boundary delimitation has highlighted the use of satellite imagery in a mapping process. Its experience in international boundary making assisted Benin and Niger in defining their boundary along the Niger River. Its work included superimposing 1:50,000 mapping from the 1960s onto recent satellite imagery. Another example where the use of satellites proved necessary was the demarcation of the Qatar-Saudi Arabia boundary. The densification of the geodetic network and reconnaissance of pillar sites was made by the same French company that complemented final documentation including maps at different scales for the catalogue of pillar coordinates.[8]

In a study on the usefulness of satellite imagery for boundary purposes, two types of data were analysed, namely ortho-images from QUICKBIRD satellites and field data, including Ground Control Points (GCPs). It was easy to identify cadastral boundaries on the satellite images that captured flat study areas. Using ortho-rectified satellite data, the boundaries of land parcels had to be digitized and the existence of vegetation further helped to easier identify the boundaries. Some of the hilly and mountainous areas were also analysed for the parcel boundaries to be extracted. The field shapes were irregular, making the analysis more difficult. Due to the mountainous terrain and irregular scattering, the satellite imagery quality was suboptimal and the shadow effect created problems in identifying boundaries efficiently.

In research conducted by the University of Durham on boundary making and territorial claims, reference was made to the impact on boundary-making processes of the availability of satellite data. Highlighting the importance of rivers as convenient natural borders, it was mentioned that the International Boundaries Research Unit (IBRU) had developed an open-access database, the International River Boundaries Database, which involved digitizing over 1,200 sections of river boundary from satellite imagery and connecting them with the information already available. This initiative resulted in a more accurate data-base on small-scale maps.[9]

---

[7] Ibid. *supra* note EOM K.S., Cartographic Section, Department of Field Support, United Nations, New York, in: OSCE, Applied Issues in International Land Boundary Delimitation/Demarcation Practices, pp. 22, 2011.

[8] Cosquer, G., (Retired) State Senior Survey Engineer, Institut Géographique National-(IGN) France, in: OSCE, Applied Issues in International Land Boundary Delimitation/Demarcation Practices, pp. 24, 2011, https://www.osce.org/secretariat/85263?download=true, (accessed 21.07.2019).

[9] University of Durham, International Boundaries Research Unit (IBRU), Boundary-making and resolving disputed territorial claims, https://impact.ref.ac.uk/casestudies/CaseStudy.aspx?Id=11842, (accessed 21.07.2019).

Satellite data as remote sensing applications are an alternative to analogue maps in the preparation phases and are also commercially available from different sources such as from SPOT and LANDSAT, QUICKBIRD. The value of satellite imagery can be assessed from various perspectives. If referring to the preparation for delimitation, then raw only data are of limited value since they require interpretation, do not carry the names of features and have no reference lines such as meridians and parallels or plane rectangular grids. However, satellite imagery is a lot more useful than analogue maps, it contains much more information such as details of agricultural cultivation, density of vegetation, water courses, temporary structures such as ditches and fences and features that are no longer in existence such as bridges, railways or buildings, and can be compared over a long period of time. For the decision phase, boundary negotiators must have all the available data, including satellite imagery, containing prominent features, which can be also verified on the ground. Such data provide decision makers with a better background for their decisions and also decreases the potential for difficulties in interpreting the intent of the signatories. "Satellite imagery, properly annotated and interpreted is an alternative to mapping, in cases when suitable scale, good quality and up to date maps are not available". Maps can be augmented by data from satellite images containing a wealth of information not available on a topographical map. After the interpretation of satellite data, a precise measurement of positions and elevations can be provided.[10]

Mapping and satellite imagery are useful for international boundary making. Supporting work to establish borders physically is important in recognition of borders. For example, the boundary between Nigeria and Cameroon was established by the 2002 judgment of the ICJ. Preliminary map sheets were prepared using satellite imagery at a scale of 1:50,000. This task was carried out by the UN Cartography Section in New York. The images were ortho-rectified, and the Digital Terrain Model (DTM) was derived from the shuttle radar topographic mission (STRM) with a 90 m resolution.[11] The Saudi-Yemeni Boundary Making Process also involved the use of satellite images used for a map at the scale of 1:1,000,000 for the desert and open areas.[12] Satellite imagery is useful in boundary delineation of agricultural fields where it can provide detailed, up-to-date, and explicit information on agricultural land use. Classification requires up-to-date field boundary maps potentially covering large areas containing thousands of farms. Automated techniques are needed for classifying crops and land uses.[13]

---

[10]Adler, R., (2000) International Boundaries Research Unit "Boundary and Territory Briefing" pp. 4–5, pp. 22, pp. 27.

[11]Sayel, M.A., Lohmann, P., Heipke, C., International Boundary Making—Three Case Studies, https://www.isprs.org/proceedings/XXXVIII/1_4_7-W5/paper/Al_Sayel-123.pdf, (accessed 23.09.2019).

[12]Ibid. *supra* note Sayel, M.A., Lohmann, P., Heipke, C., International Boundary Making—Three Case Studies, https://www.isprs.org/proceedings/XXXVIII/1_4_7-W5/paper/Al_Sayel-123.pdf, (accessed 23.09.2019).

[13]North, H.C., Pairman, D., Beliss S.E., Boundary Delineation of Agricultural Fields in Multitemporal Satellite Imagery, 2018, in: IEEE Journal of selected topics in Applied

2.2 Using Satellite Data to Defend Minorities Rights in Aid of Indigenous Communities

The International Work Group for Indigenous Affairs (IWGIA) provides various support for indigenous people.[14] The common ground is environmental protection for indigenous communities as each violation in deforestation, mining and water could be fatal. Also of concern is forced displacement due to cattle ranching and agricultural development such as soy farming. In an effort to protect their forests, indigenous communities could make use of satellite data and digital mapping technology to map the extent and legal status of the land and resources. Through the use of satellite data, indigenous communities can locate their ancestral territories and identify which lands have already been legalized and which territories they still claim. Also, such data would be useful in forest conservation. In this regard, mapping requires standardization of data collection and clear protocols.[15] Already there are publicly accessible interactive platforms, such projects could substantiate territorial claims and prevent transgressions. IWGIA was launched in November 2017, but the challenge is to evidence disputed land claims.[16] At the UN Permanent Forum on Indigenous Issues held 16–27 April 2018 at the UN Headquarters in New York, it was underlined that in many jurisdictions law enforcement is inadequate or non-existent.[17] On this occasion, the challenges for indigenous land, territories and resources were debated and it was concluded that those who defend indigenous rights have been targeted and will continue to be so.

The potential use of satellite imagery for indigenous communities includes data availability, measurement and monitoring capacity, technical assistance especially for satellite data access, data storage, data analysis/validation, and data dissemination.[18] Satellite data have been used and could be of further help to identify isolated indigenous populations. In many cases, to remain isolated means to become extinct,

Earth Observations and Remote Sensing, pp. 99: 1–15. https://doi.org/10.1109/jstars.2018.2884513, https://www.researchgate.net/publication/329817494_Boundary_Delineation_of_Agricultural_Fields_in_Multitemporal_Satellite_Imagery, (accessed 20.07.2019).

[14] IWGIA, About us, https://www.iwgia.org/en/about, (accessed 26.09.2019).

[15] Pedris, L, Tech and collaboration are putting indigenous land rights on the map, 26 March 2018, https://news.mongabay.com/2018/03/tech-and-collaboration-are-putting-indigenous-land-rights-on-the-map/, (accessed 29.06.2019).

[16] Ibid. *supra* note Pedris, L, Tech and collaboration are putting indigenous land rights on the map, 26 March 2018, https://news.mongabay.com/2018/03/tech-and-collaboration-are-putting-indigenous-land-rights-on-the-map/, (accessed 29.06.2019).

[17] UN News, Protect indigenous people's land rights and the whole world will benefit, UN forum declares. 17 April 2018, https://www.un.org/development/desa/en/news/social/unpfii17-opening.html, (accessed 29.07.2019).

[18] Riamit, S., Tauli-Corpuz, V., Indigenous Peoples' Perspectives and Activities in Monitoring Reporting, and Indicators Development for REDD+ and A Review of the MRV Concepts, Tools and Instruments, pp. 26, https://www.forestcarbonpartnership.org/system/files/documents/Indigenous%20Peoples%20Perspective%20on%20Community%20based%20MRV%20for%20Social%20and%20Enviromental%20Standards.pdf, (accessed 29.07.2019).

and the research conducted in 2016 by Walker, Kesler and Hill indicated that indigenous communities are headed towards extinction.[19] The study used 30 m resolution LANDSAT and 1 km resolution Moderate Resolution Imaging Spectroradiometer (MODIS) satellite imagery to track deforestation and identify forest clearings. The imagery was further assessed and validated with high-resolution imagery. The analysis offered data to track isolated populations through time, revealing that

> only a single example of a relatively large and growing metapopulation (cs. 50 cleared ha and 400 people), whereas all of the other 7 groups appear critically endangered with much smaller villages and gardens, and no sizable growth over time.[20]

2.2.1 *Brazil*

According to IWGIA, approximately 900,000 indigenous persons distributed among 305 ethnic groups live in Brazil. The main threats for the indigenous communities pertain to their territories. Indigenous communities are affected directly or indirectly by hydroelectric and mining companies. A serious deterioration of human rights and indigenous conditions was recently reported.[21] IWGIA showed that according to an analysis performed by the Institute of Man and Environment of the Amazon (IMAZON) 18% of the deforestation in 2018 took place in indigenous lands or state and federal protected areas, and 65% occurred in illegally occupied public areas and private lands.[22] This analysis was performed using satellite data.

2.2.2 *Costa Rica*

According to IWGIA, Costa Rica is the host country for 24 indigenous territories, comprising 6.7% of the country's land mass with approximately 100,000 persons recognized as indigenous, constituting 2.4% of the Costa Rican population.[23] Within these territories, "conflicts arise from the occupation of lands by third parties and

[19] Walker, R.S., Kesler, D.C., and Hill, R.K., Are Isolated Indigenous Populations Headed toward Extinction? March 2018, https://www.ncbi.nlm.nih.gov/pmc/articles/PMC4783021/, (accessed 29.07.2019).

[20] Ibid. *supra* note Walker, R.S., Kesler, D.C., and Hill, R.K., Are Isolated Indigenous Populations Headed toward Extinction? March 2018, https://www.ncbi.nlm.nih.gov/pmc/articles/PMC4783021/, (accessed 29.07.2019).

[21] IWGIA, Serious Deterioration of Human Rights and Indigenous Rights Conditions in Brazil, 14 March 2019, https://www.iwgia.org/en/brazil/3320-serious-deterioration-of-human-rights-and-indigenous-rights-conditions-in-brazil, (accessed 23.07.2019).

[22] Ibid. *supra* note IWGIA, Serious Deterioration of Human Rights and Indigenous Rights Conditions in Brazil, 14 March 2019.

[23] Berger D.N. (Eds.), "The Indigenous World 2019". April 2019 https://www.iwgia.org/en/documents-and-publications/documents/4-the-indigenous-world-2019/file, (accessed 23.07.2019).

illegal extraction of natural resources (lumber, animals and water, for example)". There is inadequate protection of territorial rights. The IWGIA has also reported on the so called "invasion of lands" which has affected indigenous production systems. Three indigenous Consultation Territorial Consultation Bodies were created in a pilot program by the Office of the UN High Commissioner for Human Rights (UNHCR). The new settlers turned the rainforests into pastures for cattle. Apparently, according to IWGIA, the state has not only ignored the problem, but also aggravated it by recognizing and enrolling non-indigenous foreigners as indigenous persons so that they can occupy lands.[24] Access to justice is available and indigenous people have priority for translators. Overall, discrimination persists, the territorial issue continues to be a problem and social exclusion continues to be a serious concern accentuated by deficient water services.

2.2.3 *Ecuador*

According to IWGIA, Ecuador has 14 indigenous nationalities with a population of around 1.1 million. 24.1% of the indigenous population lives in the Amazon Region and belongs to ten nationalities.[25] A referendum was held in February 2018 and underlined the challenges for the indigenous population. The referendum was not authorized by the Constitutional Court.[26] Even so, the outcome of this referendum underlined the demands of indigenous and environmental organizations. The first was "a prohibition against mining for metals in protected areas, untouchable zones and urban centres, as well as a 50,000 hectare extension to the Yasuni National Park (…)". The second demand was "a reduction from 1,300 to 300 hectares of the hydrocarbons operations area in that zone".[27] The indigenous population expressed their demands in different ways, including a march with almost 300 people advocating environmental, educational, political, anti-corruption, and communications issues. The manifesto included demands for urgent reforms to the Law on Waters, Lands, Mining, the Integral Organic Criminal Code (COIP) and to other instruments that allow the monopolization of natural resources, strip away rights and criminalize social protest.[28]

The indigenous communities, such as the A'I Cofán community of Sinangoe where approximately 180 people are located along the banks of the Aguarico River, north of the Amazon River, mostly depend on activities such as fishing, hunting and the cultivation of small family gardens. Threats to indigenous communities are principally hydrocarbons exploitation. The monitoring of the areas also revealed "illegal mining, hunting poachers, illegal logging of the forest and nonconventional fishing"

---

[24]Ibid. *supra* note Berger D.N. (Eds.), "The Indigenous World 2019". April 2019, pp. 87.

[25]Ibid. *supra* note Berger D.N. (Eds.), "The Indigenous World 2019". April 2019, pp. 161.

[26]Ibid. *supra* note Berger D.N. (Eds.), "The Indigenous World 2019". April 2019, pp. 162.

[27]Ibid. *supra* note Berger D.N. (Eds.), "The Indigenous World 2019". April 2019, pp. 163.

[28]Ibid. *supra* note Berger D.N. (Eds.), "The Indigenous World 2019". April 2019, pp. 164.

that affect the lifestyle and negatively impact the survival of the A'I Cofán community of Sinangoe.[29] The community received a positive court decision that suspended the extractive mining activities and the mining concessions.[30] The president of the Cofan community stated:

> We, the Cofán, depend upon those rivers for our lives. If the water of those rivers gets contaminated, that contamination reaches us directly, since we fish from, bathe in, and drink directly from the river.

2.2.4 *Peru*

According to IWGIA, 4 million indigenous persons live in Peru, with the largest indigenous nationality being Quechua (83.11%), followed by Aymara (10.92%). Another particularity for Peru is that 21% of the territory consists of mining concessions. Also, 75% of the Peruvian Amazon Region is covered by oil and gas concessions. The IWGIA has also stated that "defending human rights is a high-risk activity in Peru, as borne out by the figures, which indicate that 87 defenders have been killed in the country since 2011".[31] Illegal mining and human trafficking are among the challenges the indigenous people have to face. There has also been discussion of criminalization of protests. Human rights are affected by a weak democracy:

> the weaker a country's democracy the greater the vulnerability of HRDs because they are the people who denounce bad practices, expose people and situations involving abuses of power and defend the basic rights of individuals.[32]

Development of a mechanism to protect human rights has been discussed in Peru and various campaigns have supported the work of individuals and organizations for human rights. Deforestation is a growing problem for Peru and the whole of the Amazon Region, "being caused by a proliferation of illegal mining and logging mafias".[33] Indigenous communities are affected and are looking for solutions in their struggle against deforestation. Based on NGO data, the figures for deforestation in Peru presented by IWGIA are alarming, some 150,000 hectares were calculated as the annual deforestation. To fight deforestation, the so-called "Land Security for Indigenous People" programme implemented through the Peruvian Environmental Rights Society supports the creation of brigades of native community members with

---

[29] Ibid. *supra* note Berger D.N. (Eds.), "The Indigenous World 2019". April 2019, pp. 166–167.

[30] Case No. 21333201800266, Provincial Court of Sucumbios, October 2018

See also: UN GA A/HRC/42/37/Add.1, Report of the Special Rapporteur on the rights of indigenous people, Paragraph 41, 4 July 2019, http://unsr.vtaulicorpuz.org/site/images/docs/country/2019-ecuador-a-hrc-42-37-add1-en.pdf, (accessed 26.09.2019).

[31] Ibid. *supra* note Berger D.N. (Eds.), "The Indigenous World 2019". April 2019, pp. 198.

[32] Ibid. *supra* note Berger D.N. (Eds.), "The Indigenous World 2019". April 2019, pp. 200.

[33] Ibid. *supra* note Berger D.N. (Eds.), "The Indigenous World 2019". April 2019, pp. 201.

the aim for "establishing boundary markers as points by which to geo-reference their territories in the face of the advancing mafias".[34]

The lack of a law on territorial organization means that peasant communities do not have private property on land—it is all communally-owned, the sale of land being possible only with the approval of two-thirds of the members of that community.[35] Within this framework problems have arisen, with group of people trying to sell the land in communal use, resulting in increased cases of the "crime of land grabbing" in the Amazon Region. For the environment, serious problems are posed by the extraction projects in Peru, in particular where fracking as a method of fossil fuel exploration is proposed. Past experiences in Peru show the negative impact on the land of such exploitation causing "serious pollution of the Marañón, Tigre, Corrientes and Amazonas river basins around Plot 192", for example.[36]

2.3 Satellite Data and Ethnic Conflicts

Satellite imagery has been useful in documenting the violent removal of entire villages (and populations) in rural population/re-settlement (in Africa), including land issues, and expropriation of the rural population for industry and investors. The following cases will showcase the use of satellite imagery to prove violations of human rights.

2.3.1 Ethiopia

Various examples prove that satellite imagining has become necessary in certain cases to prove infringement of the law and violations of human rights. By using satellite images, Human Rights Watch showed the forced displacement and violations of basic human rights in the process of villagization in Ethiopia's Gambella Region. Allegedly, Ethiopian Government policy was harmful for the area's rural population by "forcibly moving tens of thousands of indigenous people in the western Gambella region from their homes to new villages" without assuring the necessary conditions to survive, threatening their access, and right, to basic services.[37] Human Rights Watch found widespread human rights violations after interviewing over 100 residents, over a four-week period in Ethiopia, from May to June 2011, affected in the

[34] Ibid. *supra* note Berger D.N. (Eds.), "The Indigenous World 2019". April 2019, pp. 201.

[35] Ibid. *supra* note Berger D.N. (Eds.), "The Indigenous World 2019". April 2019, pp. 203. See also analysis of Juan Carlos Ruiz Molleda referenced at pp. 203.

[36] Ibid. *supra* note Berger D.N. (Eds.), "The Indigenous World 2019". April 2019, pp. 204.

[37] Human Rights Watch News, Waiting Here for Death. Forced Displacement and Villagization in Ethiopia's Gambella Region, 16 January 2012, https://www.hrw.org/report/2012/01/16/waiting-here-death/forced-displacement-and-villagization-ethiopias-gambella-region#page, (accessed 10.05.2019).

villagization program. The general problem was that from 2008 Ethiopia started to lease huge portions of land to foreign and domestic investors without meaningful consultations or compensation for the indigenous people of these areas to be relocated. The governmental program promised "socioeconomic and cultural transformation" but delivered the opposite, "doing more to undermine rights and livelihoods of the population than to improve them".[38] The forced relocation brought hunger and abuse of the indigenous population.[39]

The mass displacement was hard to prove, research and monitor from the field, due to various factors.[40] In general, procedures to locate, identify and interview individuals are not easy in conflict zones or in other areas with high-risks for foreign scientists or reporters. To minimize the likelihood of terrorist attacks or other unlawful actions to the foreign interviewers or the safety and security of the interviewee, to have instant access to information and real time images or access to previous images, satellite technology can be used to support the work in the field in certain areas and periods and document remote locations where the risk of conflict is high.

To evidence the rural displacement and villagization progress, Human Rights Watch has used satellite images provided by the American Association for the Advancement of Science (AAAS) "Geospatial Technologies and Human Rights" to prove infringement of the indigenous rights. "The satellite images are a powerful tool for illustrating these violations. They provide irrefutable evidence".[41] Satellite images have confirmed that villagers have been relocated to plots of land of worse quality than that from where they were moved, possibly to make way for large foreign-owned commercial farms.[42] The AAAS report revealed that significant changes took place in the spatial distribution of structures and agriculture between 2007 and 2011.[43] The rural population was not consolidating around the modern infrastructure but in fact started to disappear.

---

[38] Ibid. *supra* note HRW Waiting Here for Death. Forced Displacement and Villagization in Ethiopia's Gambella Region, 16 January 2012.

[39] Human Rights Watch Press Release, Ethiopia: Forced Relocations Bring Hunger, Hardship. Donors Fund Should not Facilitate Abuse of Indigenous Groups, 16 January 2012, https://www.hrw.org/news/2012/01/16/ethiopia-forced-relocations-bring-hunger-hardship, (accessed 10.05.2019).

[40] Ibid. *supra* note HRW Waiting Here for Death. Forced Displacement and Villagization in Ethiopia's Gambella Region, 16 January 2012.

> Laws that severely infringe on the functioning of NGOs including the Charities and Societies Proclamation and the Anti-Terrorism Proclamation; restrictions on media freedoms; the government's intolerance of political dissent; and the intimidation and fear generated by government officials that permeates life in Ethiopia.

[41] Maxmen, A., AAAS Geospatial Report: Ethiopian "Villagization" Policy is Displacing Farmers in Gambella Region, 17 January 2012.

See also: Froehlich, A., Space Applications Supporting Justice, 2019 in: Froehlich, A., (Eds.) Embedding Space In African Society. The United Nations Sustainable Development Goals 2030 Supported by Space Applications.

[42] Ibid. *supra* note Maxmen, A., AAAS Geospatial Report: Ethiopian "Villagization" Policy is Displacing Farmers in Gambella Region, 17 January 2012.

[43] AAAS Report "Documentation of Villagization: Gambella Region, Ethiopia".

In this particular case, the result of using satellite imagery was equal to the research in the field. "Using satellite imagery, we came up with the same result as people on the ground" said Susan Wolfinbarger, Senior Program Associate for the AAAS Geospatial Technologies and Human Rights Project. The combination of modern tools such as satellite images, geographic information/EO systems and GNSS are a useful tool that is increasingly necessary to support the work of local and international human rights organizations. The most important feature of satellite imaging is the capability to compare images of the same location but taken in different temporal periods. For example, in the case of the Ethiopian villagization, a satellite photo taken in 2009 revealed a community with 68 structures near cultivated fields while in the same location in 2011 the structures had disappeared and the fields lay barren, according to Jonathan Drake, analyst in the Geospatial Technologies Project.[44] On the other hand, only major transformations on land could be seen from the satellite. Smaller scale processes such as destruction of roads, or buildings are not that obvious, which is why a partnership between an organization on the ground whose work is supported by satellite imaging is to be recommended.

The Human Rights Council Working Group on the Universal Periodic Review in its Nineteenth Session from 28 April to 9 May 2014 submitted the National Report on Human Rights for Ethiopia, highlighting human rights as fundamental rights guaranteed by the Constitution.[45] The National Report stressed that the Ethiopian Human Rights Commission opened branch offices in a number of regions, including the Gambella region. The Institution of the Ombudsman and the growing number of administrative complaints from the public, recognizing the increasing number of citizens confident in this institution, was also mentioned.

In relation to the activity of the NGOs, the Report stated that the Government of Ethiopia issued the Charities and Societies Proclamation in February 2019, which "aims at strengthening the role on non-governmental organizations and enhancing their contribution in the socio-economic development of the country" which was adopted following public discussion with the NGOs.[46] The Proclamation provides a legal and conducive working environment for NGOs.

The Report also stated that Ethiopia was cooperating with the international community, IGOs and NGOs, inter alia, UNHCR, WFP, and UNICEF, facilitating the entry of such organizations into the country, including the areas where there are

---

See also: AAAS Human Rights Applications of Remote Sensing. Case Studies from the Geospatial Technologies and Human Rights Project, 2014, https://www.aaas.org/sites/default/files/reports/Human_Rights_Applications_of_Remote_Sensing.pdf, (accessed 23.09.2019).

[44]Ibid. *supra* note Maxmen, A., AAAS Geospatial Report: Ethiopian "Villagization" Policy is Displacing Farmers in Gambella Region, 17 January 2012.

[45]"*The Constitution of the Federal Democratic Republic of Ethiopia (FDRE) is the foundation of the human rights protection and the basis to build the country's democratic system. The Constitution, under Article 9(4) provides that international agreements ratified by Ethiopia form an integral part of the law of the land and under Article 13 enshrines the responsibility of all branches of Government to respect and ensure respect the fundamental rights and freedoms*".

[46]UNGA A/HRC/WG.6/19/ETH/1, (2014). "*National report submitted in accordance with paragraph 5 of the annex to Human Rights Council resolution 16/21 Ethiopia*", Paragraph 27.

conflicts.[47] The Report from 2014 concluded with Ethiopia's request for technical assistance necessary for implementing the human rights instruments ratified by Ethiopia and their dissemination.[48] Included among the challenges in the promotion and protection of human rights in Ethiopia were "poverty, resource constraints, lack of capacity, and climate change impacts" which represent factors that make it "difficult to create uniform strategies and programs for the promotion and protection of human rights in all regions".[49]

The Ethiopian Government Villagization Program was also discussed at the UN. UNESCO stated that there were marginalized communities (minorities and indigenous people) unable to exercise their right to take part in cultural life. The Special Rapporteur on the Rights of Indigenous People received claims in 2012 about the Ethiopian Government Villagization Program.[50] As stated at the UN, the program consisted of the "relocation of pastoralists and agro-pastoralists and the shifting of cultivators into sedentary villages" where supposedly they would have been provided with improved social services, housing and infrastructure.[51] However, numerous concerns were raised including that the government had failed in obtaining the consent of affected indigenous groups and the lack of services at the resettlement sites.[52]

2.3.2 Report of the Independent International Fact-Finding Mission on Myanmar

The independent international fact-finding mission on Myanmar was established by the Human Rights Council in its resolution 34/22. Since 2011, the mandate has been to investigate the infringement of fundamental freedoms. The conclusions have established consistent patterns of human rights violations and abuses of international humanitarian law. As part of the methodology, the factual finding was based on the "reasonable grounds" standard of proof, meaning it was necessary to gather evidence that would allow an ordinary person to reasonably conclude that an incident or pattern

---

[47] Ibid. *supra* note UNGA A/HRC/WG.6/19/ETH/1, (2014), Paragraph 101.

[48] Ibid. *supra* note UNGA A/HRC/WG.6/19/ETH/1, (2014), Paragraph 109.

[49] Ibid. *supra* note UNGA A/HRC/WG.6/19/ETH/1, (2014), Paragraph 108.

[50] UNGA A/HRC/WG.6/19/ETH/2, (2014). "*Compilation prepared by the Office of the United Nations High Commissioner for Human Rights in accordance with paragraph 15 (b) of the annex to Human Rights Council resolution 5/1 and paragraph 5 of the annex to Council resolution 16/21*", Paragraph 69,
https://documents-dds-ny.un.org/doc/UNDOC/GEN/G14/109/38/PDF/G1410938.pdf?OpenElement, (accessed 11.05.2019).

[51] Ibid. *supra* note UNGA A/HRC/WG.6/19/ETH/2, (2014), Paragraph 69.

[52] UNGA A/HRC/22/67, (2013). "*Communications report of Special Procedures*", pp. 126, https://www.ohchr.org/Documents/HRBodies/HRCouncil/RegularSession/Session22/A-HRC-22-67_EFS.pdf, (accessed 11.05.2019).

of conduct occurred.[53] Satellite imagery contributed to the vast amount of primary information of the mission, the Commission mentioning that the mission relied only on verified and corroborated information.[54] The procedure of gathering evidence adhered to the principles of independence, impartiality and objectivity with special attention to protecting human rights defenders.[55]

In particular, in 2018 the Commission report underlined that satellite imagery and first-hand accounts corroborated widespread, deliberate and targeted destruction, with more than 393 villages partially or totally destroyed in northern Rakkine.[56] Satellite imagery supported evidence of partial or totally destruction of a wide area for investigation, totalling 40 percent of all settlements, which the Commission acknowledged encompassed at least 37,700 individual structures. Satellite imagery was particularly helpful in gathering evidence because of the attempts to destroy criminal evidence of the violation of human rights in Rohingya. In fact, attempts were made to erase the mass displacement and destruction and burning of villages and settlements in Rohingya. By gathering evidence with the help of satellite imagery the Commission evidenced that "The mass displacement and the burning of Rohingya villages were followed by the systematic appropriation of the vacated land. Bulldozers flattened burned, damaged and even surviving structures and vegetation, erasing every trace of the Rohingya communities, while also destroying criminal evidence (…)".[57] The impunity of the security forces that perpetrated in gross human rights caused the exodus of the citizens of Myanmar and the evidence gathered by the Commission was to make them responsible in front of justice. Even though the Burmese Government made an in principle commitment to the repatriation of citizens in Rohingya, the fact that the same security forces that were the perpetrators of human rights violations were to be responsible for ensuring the security of returnees, raised questions as to the efficacy of such measures.[58]

On the basis of the body of information collected by the Commission, including the analysis of satellite imagery for wide and remote areas, the mission concluded that it "has reasonable grounds to conclude that serious crimes under international law have been committed that warrant criminal investigation and prosecution". Fact pointed out by the evidence gathering indicated elements of genocide, in particular "broader oppressive context and hate rhetoric, utterances of commanders and direct

53 A/HRC/39/64, Report of the Independent international fact-finding mission on Myanmar, Paragraph 6, 2018, https://www.ohchr.org/EN/HRBodies/HRC/MyanmarFFM/Pages/ReportoftheMyanmarFFM.aspx, (accessed 05.06.2019).

54 Ibid. *supra* note A/HRC/39/64, Report of the Independent international fact-finding mission on Myanmar, Paragraph 7, 2018.

55 Ibid. *supra* note A/HRC/39/64, Report of the Independent international fact-finding mission on Myanmar, Paragraph 9, 2018.

56 Ibid. *supra* note A/HRC/39/64, Report of the Independent international fact-finding mission on Myanmar, Paragraph 42, 2018.

57 Ibid. *supra* note A/HRC/39/64, Report of the Independent international fact-finding mission on Myanmar, Paragraph 50, 2018.

58 Ibid. *supra* note A/HRC/39/64, Report of the Independent international fact-finding mission on Myanmar, Paragraph 51, 2018.

perpetrators, exclusionary policies including altering the demographic composition of Rakhine State" which lead to the intention to destroy, in whole or in part, a national, ethnical, racial or religious group as such.[59] The extreme scale and brutality of the violence required careful consideration by the Commission which concluded that there was sufficient information to warrant the investigation and prosecution of senior officials and the need for a competent court to determine their liability for genocide in relation to the situation in Rakhine State.[60]

In addition to genocide, crimes against humanity were evidenced by the Commission, referring to murder, imprisonment, enforced disappearance, torture, rape, sexual slavery, sexual violence, persecution and enslavement.[61] Also, the evidence gathered by the Commission fulfilled the elements describing war crimes. In particular, among the gathered evidence, the Commission acknowledged "elements of murder, torture, cruel treatment, outrages upon personal dignity attacking civilians, displacing civilian, pillaging, attacking protected objects, taking hostages, sentencing or execution without due process".[62]

Overall, the perpetrators had not been held accountable mainly due to authorities that systematically failed to condemn, investigate and prosecute perpetrators.[63] The Commission acknowledged that none of the efforts to create ad hoc inquiry commissions and boards in Myanmar had met the standard of impartial, independent, effective and thorough human rights investigation.[64] In addition, the impunity of security forces threatened the political and legal system, placing the perpetrators above the law.[65] For this reason, the Commission concluded that the Government of Myanmar "is unable and unwilling" to investigate and prosecute crimes under international law and that the impetus for accountability should be led from the international community.[66] This is the reason why, in such complicated and dangerous social and political context, satellite imagery is an essential element to prove violations of human rights before an international or even a national court of justice.

The recommendation made by the Commission to the Human Rights Council was to highlight the need and create an "independent, impartial mechanism to collect,

[59] Ibid. *supra* note A/HRC/39/64, Report of the Independent international fact-finding mission on Myanmar, Paragraph 85, 2018.

[60] Ibid. *supra* note A/HRC/39/64, Report of the Independent international fact-finding mission on Myanmar, Paragraph 87, 2018.

[61] Ibid. *supra* note A/HRC/39/64, Report of the Independent international fact-finding mission on Myanmar, Paragraph 88, 2018.

[62] Ibid. *supra* note A/HRC/39/64, Report of the Independent international fact-finding mission on Myanmar, Paragraph 89, 2018.

[63] Ibid. *supra* note A/HRC/39/64, Report of the Independent international fact-finding mission on Myanmar, Paragraph 95, 2018.

[64] Ibid. *supra* note A/HRC/39/64, Report of the Independent international fact-finding mission on Myanmar, Paragraph 96, 2018.

[65] Ibid. *supra* note A/HRC/39/64, Report of the Independent international fact-finding mission on Myanmar, Paragraph 97, 2018.

[66] Ibid. *supra* note A/HRC/39/64, Report of the Independent international fact-finding mission on Myanmar, Paragraph 98, 2018.

consolidate, preserve and analyse evidence of violations of international humanitarian law and human rights violations and abuses and to prepare files to facilitate and expedite fair and independent criminal proceedings in national, regional or international courts or tribunals".[67] Such measures were expected before the UN Security Council took action.

2.4 Human Rights Related Satellite Data in Wartime

The following cases illustrate the use of satellite imagery to prove violations of human rights in wartime. Cases will cover the dispute between the U.S. and Iraq in the Oil Platform Case, the Bosnian mass graves, the use of Barrel Bombs in Syria, the monitoring of refugees in crisis areas, an attack on UN camps, and illicit arms traffic.

2.4.1 The Oil Platforms Case

In 1992 the Islamic Republic of Iran (Iran) filled an application against the USA in regard to the destruction of Iranian oil platforms.[68] In its counter-memorial, the U.S. requested the ICJ to also decide that Iran breached its obligations by attacking vessels in the Persian Gulf and engaging in military actions detrimental to commerce and navigation.[69]

The evidence produced by the U.S. included images, taken by satellite or aerial reconnaissance aircraft, of the FAW area and of the four alleged missile sites under Iranian control at the time of the attack, as well as a complementary expert report describing and examining this imagery.[70] Iran suggested that no credible evidence had been produced that there were operational Iranian missile sites in the FAW area; Iran stated that the satellite images produced by the U.S. were not very clear, and appealed to its own experts' opinion to prove that the installations shown therein "bear no resemblance to a normal Silkworm missile site".[71] The U.S. claimed that

[67] Ibid. supra note A/HRC/39/64, Report of the Independent international fact-finding mission on Myanmar, Paragraph 107, 2018.

[68] ICJ, Memorial submitted by the Islamic Republic of Iran, 8 June 1993, https://www.icj-cij.org/files/case-related/90/8622.pdf, (accessed 26.09.2019).

[69] ICJ, Counter-Memorial and Counter-Claim submitted by the United States of America, 23 June 1997, https://www.icj-cij.org/files/case-related/90/8632.pdf, (accessed 26.09.2019).

[70] International Court of Justice, (2003) Oil Platforms (Iran v. U.S.), 2003 I.C.J. 161 (Nov. 6), *Oil Platforms: Iran vs. United States Judgement,* 6 November 2003, General List No. 90, Paragraph 53 http://www.worldcourts.com/icj/eng/decisions/2003.11.06_oil_platforms.htm, (accessed 13.05.2019).

[71] Ibid. *supra* note International Court of Justice, (2003) Oil Platforms (Iran v. U.S.), 2003 I.C.J. 161 (Nov. 6) *Oil Platforms: Iran vs. United States Judgement*, 6 November 2003, Paragraph 54.

its satellite imagery showed that there was no Iraqi missile launching facility in the FAW area at the time. It also affirmed, on the basis of an independent expert's opinion, that Iraq's HY-2 missiles were not equipped with a system capable of guiding them along a circuitous path, as contended by Iran. Finally, the U.S. rejected the Iranian theory that the missile was launched from air or sea.[72] As noted above, the United States claimed that the missile that struck the Sea Isle City was a ground-launched HY-2 anti-ship missile of the type known as the "Silkworm", but it was not able to produce physical evidence of this, for example in the form of recovered fragments of the missile. The ICJ examined other evidence on the hypothesis that the missile was of this type. The U.S. also claimed that the missile was fired from Iranian-held territory in the FAW area, and it offered satellite pictures and expert evidence to show that there was, at the time, Iranian missile-firing equipment present there. Even with the assistance of the expert reports offered by both parties, the ICJ did not find the satellite images sufficiently clear to establish this point.[73]

The use of satellite imagery as evidence was crucial in the "Oil Platforms" case. In supporting its argument that Iran was using off shore oil platforms as missile sites to carry out attacks against neutral ships during the Iran-Iraq war, the U.S. provided satellite imagery of the FAW area and another four alleged missiles sites under Iranian control at the time of the attack. Iran claimed that the satellite images in fact proved that the installations bore no resemblance to a normal Silkworm missile site. The U.S. claimed that this evidence indicated an Iranian land-launched missile. The ICJ did not however find the satellite images to be sufficiently clear to decide that Iranian missile-firing equipment was present there, especially if analysed together with the testimony from the Kuwaiti military officer, which was given ten years after the reported incidents and with a discrepancy between the English and the Arabic version which "lacks any indication of the bearing on which the observed missile was travelling". The ICJ also found that the U.S. presented its claim in a generic sense while it was necessary to prove an actual impediment to commerce or navigation between the territories of the two High Contracting Parties. Thus, the ICJ concluded that there was no need for it to consider under this case the contested issues of attribution of those incidents to Iran. The ICJ found that the U.S. had not breached its obligations and rejected Iran's claim. Also, the Court found that there was no impediment to commerce or navigation between the territories of Iran and the U.S., and rejected the U.S. counter-claim for reparation.

[72]Ibid. *supra* note International Court of Justice, (2003) Oil Platforms (Iran v. U.S.), 2003 I.C.J. 161 (Nov. 6), *Oil Platforms: Iran vs. United States Judgement*, 6 November 2003, Paragraph 56.

[73]Ibid. *supra* note International Court of Justice, (2003) Oil Platforms (Iran v. U.S.), 2003 I.C.J. 161 (Nov. 6), *Oil Platforms: Iran vs. United States Judgement*, 6 November 2003, Paragraph 58.

2.4.2 *Bosnian Mass Graves*

Satellite images were the first to be used as evidence by an international court, the International Tribunal for the former Yugoslavia (ICTY) to prosecute Bosnian Serb generals Kristic and Mladic.[74] The ICTY was established pursuant to Security Council Resolution 827 (1993) "for the sole purpose of prosecuting persons responsible for serious violations of international humanitarian law committed in the territory of the former Yugoslavia" from 1 January 1991. The creation of ICTY represented the crystallization of international humanitarian law in the Geneva Conventions of 1949 and the Genocide Convention of 1948. The satellite imagery represented important evidence and supported the procedure to prove guilt beyond reasonable doubt.[75]

Mass graves with nearly 100 victims from Bosnia's civil war were discovered in the mountainous area of Koricanske Stijene in central Bosnia. The massacre was one of the most brutal episodes during the events of the 1992–1995 inter-ethnic war in Bosnia. Various exhumations in the area revealed other mass graves. The remains of 25,000 victims have now been exhumed from hundreds of mass graves.[76] Filled with the bodies of dozens of victims, mass graves were very hard to follow. The Washington representatives showed at the Security Council aerial photos and satellite imagery with freshly disturbed earth. In the news, the photographs from spy satellites and U-2 military reconnaissance planes showed a long curving road by a football pitch, with areas that had been freshly dug up. This was the only way to prove violations of human rights as access to those areas was not allowed to UN officials.[77] The mass grave near Prijedor was considered to be the largest mass grave in Bosnia and Herzegovina.

Satellite images were used extensively by the Bosnian prosecutors and the International Commission on Missing Persons to find the undiscovered mass graves from the 1990s conflict. It was stated that the satellite imagery was used due to the need to speed up the process but also because of the difficulties in finding and identifying the missing.[78] The International Commission on Missing Persons (ICMP) used a multi-disciplinary team of researchers to detect the geographic patterns associated with mass graves. The multi-disciplinary team of satellite imagery experts, geology

[74] UN International Criminal Tribunal for the Former Yugoslavia, History, http://www.icty.org/en/about/office-of-the-prosecutor/history, (accessed 23.09.2019).

[75] United Nations ICTY, 20 years of the ICTY Anniversary Events and Legacy Conference Proceedings, 2014, http://www.icty.org/x/file/Press/Events/2013/20_Years_of_the_ICTY_en.pdf, (accessed 23.09.2019).

[76] The Telegraph, Mass grave with nearly 100 victims found in Bosnia, 21 September 2017, https://www.telegraph.co.uk/news/2017/09/21/mass-grave-nearly-100-victims-found-bosnia/, (accessed 06.08.2019).

[77] Tran, M., The Guardian, Spy pictures show Bosnia massacre, 11 August 2019, https://www.theguardian.com/world/1995/aug/11/warcrimes.marktran, (accessed 06.08.2019).

[78] Dzidic, D., Bosnia to Hunt Mass Graves Using Satellite Images, 27 March 2013, https://balkaninsight.com/2013/03/27/bosnia-will-use-satellite-imagery-to-locate-mass-graves/, (accessed 23.09.2019).

experts and forensic archaeologists were to find the mass graves in Bosnia.[79] Mass graves were also found in Srebrenica in 2015, containing remains that were most likely victims of the 1995 Srebrenica genocide.

Investigations under the ICTY included extensive witness interviews and analysis of U.S. satellite imagery that showed large areas of freshly dug soil on the dates when the executions were occurring, which helped the prosecutors and the Tribunal to identify a number of locations suspected to contain mass graves. Also, satellite imagery supported the work of the investigators and led them to alleged killing sites.[80]

Transcripts from the ICTY prove the usefulness of the satellite imagery provided by the U.S.. The content of the transcripts however refers to aerial imagery even if the imagery was mostly generated by remote sensing satellites and not only by planes. For example, the Petkovici Dam primary grave to the Liplje secondary graves used remote sensing imagery to indicate that the primary grave was dug between 5 and 27 July 1995.[81]

In another case, it was underlined that a number of photographs usually referred to as aerial images was provided to Prosecution as part of the imagery from the U.S. intelligence community. Such evidence included restrictions in regards to the technical means:

> the Prosecution may only refer to the "U.S. reconnaissance systems" as the author of the material, but the Prosecution is not authorised to "discuss in courtroom proceedings any information relating to the technical or analytical sources, methods, or capabilities of the systems, organisations, or personnel used to collect, analyse, or produce these imagery-derived products," and the "Trial Chamber may not order either party to produce additional evidence received from the person or entity providing the initial information.

For this reason, in the Final Trial Brief, the Defence underlined that even if the imagery may not have appeared to provide accurate, objective information about a particular location, the fact that no evidence or explanations were presented to the Trial Chambers as to whether these are satellite photographs, photographs taken by unmanned aircraft or photographs taken by freezing the frames of a video recording, the Chamber may not rely on such imagery when rendering the final judgement. It was recalled that in order to convert the raw data into a useful visualization, the images to the Court had to be interpreted and converted from raw data that may be exaggerated or biased depictions of the underlying data.[82]

---

[79] ICMP Press Release, ICMP Finds Improved Methods for Locating Mass Graves, https://www.icmp.int/press-releases/icmp-finds-improved-methods-for-locating-mass-graves/, (accessed 23.09.2019).

[80] UN International Criminal Tribunal for the former Yugoslavia, Investigations, http://www.icty.org/en/content/investigations-0, (accessed 23.09.2019).

[81] The International Criminal Tribunal for the Former Yugoslavia, Case No. IT-05-88/2-PT, 29 September 2009, http://www.icty.org/x/cases/tolimir/custom3/en/090929.pdf, (accessed 23.09.2019).

[82] International Criminal Tribunal for the Former Yugoslavia, Case No. IT-05-88/2-T, Para. 156–160, http://www.icty.org/x/cases/tolimir/custom5/en/121001.pdf, (accessed 23.09.2019).

2.4.3 *Use of Barrel Bombs in Syria*

The Syrian Network for Human Rights (SNHR) is an NGO founded in 2011 and based in the Syrian Arab Republic, which monitors the Syrian Civil War, in particular human right violations, and produces various statistics.[83] The SNHR Report has described the regime in Syria that has dropped nearly 70,000 barrel bombs as "Ruthless Bombing".[84] The situation in Syria has not been sent to the International Criminal Court because it was vetoed at the UN Security Council but diplomatic circles have raised the prospect of taking Syria before the world court in respect of the crimes against humanity committed by the Syrian regime.[85]

Human Rights Watch has used multiple means of information including satellite imagery, witness statements and video and photographic evidence to identify more than 400 locations in rebel-held towns and villages in Daraa, and more than 1,000 in Aleppo, where the Syrian regime has used large air-dropped munitions including the so-called barrel bombs.[86] Satellite imagery analysis has documented the attacks in Aleppo governorate in northern Syria and in Daraa governorate in the south in addition to evidence based on witness statements, video and photographs.[87] By examining satellite imagery, Human Rights Watch identified major damage in sites held by rebel groups in Daraa and in Aleppo between 22 February 2014 and 25 January 2015.

The human rights situation in the Syrian Arab Republic was also monitored by the OHCHR, which established the Independent International Commission of Inquiry (Commission) pursuant to Human Rights Council Resolution S-17/1. The scope of the Commission is:

> to investigate all alleged violations of international human rights law since March 2011 in the Syrian Arab Republic, to establish the facts and circumstances that may amount to such violations and of the crimes perpetrated and, where possible, to identify those responsible with a view to ensuring that perpetrators of violations, including those that may constitute crimes against humanity, are held accountable.[88]

---

[83] Syrian Network for Human Rights (SNHR), http://sn4hr.org, (accessed 11.05.2019).

[84] Reliefweb, The Assad Regime Has Dropped Nearly 70,000 Barrel Bombs on Syria–The Ruthless Bombing, https://reliefweb.int/report/syrian-arab-republic/assad-regime-has-dropped-nearly-70000-barrel-bombs-syria-ruthless, (accessed at 11.05.2019).

[85] Goldston, J.A., We need an ICJ ruling on Syria, 27 June 2015, https://www.opendemocracy.net/en/north-africa-west-asia/we-need-icj-ruling-on-syria/, (accessed 14.05.2019).

[86] Shaheen, K., Syria has used barrel bombs in hundreds of locations, says rights group, 25 February 2015, https://www.theguardian.com/world/2015/feb/25/syria-barrel-bombs-hundreds-locations-human-rights-watch, (accessed 23.09.2019).

[87] HRW, Syria: New Spate of Barrel Bomb Attacks, 24 February 2019, https://www.hrw.org/news/2015/02/24/syria-new-spate-barrel-bomb-attacks, (accessed 23.09.2019).

[88] A/HRC/RES/S-17/1, Resolution adopted by the Human Rights Council at its seventeenth special session. S-17/1 "Situation of human rights in the Syrian Arab Republic" (2011), Paragraph 13, https://www.ohchr.org/Documents/HRBodies/HRCouncil/CoISyria/ResS17_1.pdf, (accessed 16.05.2019).

The Commission was requested to make public the findings of the investigation and present a written update on the situation in the Syrian Arab Republic at the 17th special session of the Human Rights Council on the "Situation of Human Rights in the Syrian Arab Republic" held 22 August 2011.[89] The Human Rights Council has a web page dedicated to the Commission of Inquiry on the Syrian Arab Republic where Reports have been published since 2011 when the First Report of the Independent International Commission of Inquiry on the Syrian Arab Republic was issued and published.[90] Since beginning its work, the Commission has produced more than 20 reports and press releases. Current Commissioners come from Brazil, USA and Egypt.

When referring to the methods of work, the First Report mentioned that it had collected first-hand information through interviews with victims and witnesses of events in the Syrian Arab Republic with the aim "to investigate alleged violations of international human rights law and sought to establish the facts and circumstances that may amount to such violations".[91] Following this, the Commission underlined in 2012 the lack of access to the country which *"posed particular challenges for the documentation of abuses committed (…) given that most victims and witnesses of such abuses have remained in the country and the Government had not facilitated interviews with victims of armed group violence during the period under review"*.[92] The efforts of the Commission focused on interviewing victims and witnesses of violations, defectors and other individuals who had left the country, for which it travelled to several countries to gather testimony from people who had recently fled the Syrian Arab Republic.

Among the means to access evidence, the Commission of Inquiry expressly mentioned the usefulness of the analysis of satellite imagery of areas where military and security forces were deployed and related violations occurred, following which it "corroborated a number of witness accounts".[93] The commission explicitly stated the need to adapt its methodology in view of the challenges to physically access the country, mentioning that it had conducted interviews by telephone or Skype and applied the standard of proof used in previous reports, namely "reasonable grounds

[89] UNHRC, 17th Special Session on the "Situation of human rights in the Syrian Arab Republic", 22 August 2011, https://www.ohchr.org/EN/HRBodies/HRC/SpecialSessions/Session17/Pages/17thSpecialSession.aspx, (accessed 29.09.2019).

[90] UN Human Rights Council, Independent International Commission of Inquiry on the Syrian Arab Republic, https://www.ohchr.org/EN/HRBodies/HRC/IICISyria/Pages/Documentation.aspx, (accessed 23.09.2019).

[91] UNGA A/HRC/S-17/2/Add.1 (2011), Paragraph 10, https://documents-dds-ny.un.org/doc/UNDOC/GEN/G11/170/97/PDF/G1117097.pdf?OpenElement, (accessed 16.05.2019).

[92] UNGA A/HRC/19/69—2nd Report of the Commission of Inquiry on the Syrian Arab Republic, Paragraph 8–9, https://documents-dds-ny.un.org/doc/UNDOC/GEN/G12/106/13/PDF/G1210613.pdf?OpenElement, (accessed 17.05.2019).

[93] UNGA A/HRC/19/69, (2012). 2nd Report of the Commission of Inquiry, https://documents-dds-ny.un.org/doc/UNDOC/GEN/G12/106/13/PDF/G1210613.pdf?OpenElement, (accessed 16.05.2019).

to believe".[94] When examining the satellite imagery, the commission found that the movement of vehicles or weapons, as well as the size of the group, would have been easily detectable and not possible for any sizeable anti-government armed group.[95] This was one of the first *expressis verbis* evidence of the Commission based on satellite imagery.

In 2013, the Commission's report acknowledged the escalation in the armed conflict between the government forces and anti-government forces and was the first statement of the continuous use of barrel bombs dropped by helicopters or from fighter aircraft at high altitudes, which contributed to the violation of international humanitarian law.[96] Referring to the report Glossary of Weapons used in Syria, the so called barrel bombs consist of storage tanks or sheet-metal cylinders packed with explosives and metal scrap. It was stated that Syrian Air Forces (SAF) fighter jets and helicopters were used "extensively" during bombing of the anti-government armed groups and the areas under their influence. Barrel bombs were used indiscriminately and dropped on civilian objects causing loss of life and significant destruction of civilian property.[97] This type of weapons was used in areas of particular strategic importance such as Aleppo and Homs city.[98] The Commission continued to include access to satellite imagery as part of the necessary evidence for the investigation, in addition to photographs, video recordings and medical records together with reports from governments and NGO sources, including academic and UN reports.[99]

Detailed explanations about the use of barrel bombs increasingly employed by the SAF and the consequences to human rights, causing "devastating consequences" for civilians and civilian objects, are found in Annex VI of the 2014 Report.[100] Barrel bombs are described as highly imprecise with an extensive impact zone. Dropping

---

[94] UNGA A/HRC/21/50, Report of the Independent International Commission of Inquiry on the Syrian Arab Republic, Paragraph 9–11, 2012, https://www.ohchr.org/Documents/HRBodies/HRCouncil/RegularSession/Session21/A-HRC-21-50_en.pdf, (accessed 16.05.2019).

[95] Ibid. *supra* note UNGA A/HRC/21/50, Report of the Independent International Commission of Inquiry on the Syrian Arab Republic, Paragraph 46, 2012, https://www.ohchr.org/Documents/HRBodies/HRCouncil/RegularSession/Session21/A-HRC-21-50_en.pdf, (accessed 16.05.2019).

[96] UNGA A/HRC/22/59, Report of the Independent International Commission of Inquiry on the Syrian Arab Republic, Paragraph 6, 7, 8, 86, 90–91, 2013, https://www.ohchr.org/Documents/HRBodies/HRCouncil/CoISyria/A.HRC.22.59_en.pdf, (accessed 26.09.2019).

[97] Ibid. *supra* note UNGA A/HRC/22/59, Report of the Independent International Commission of Inquiry on the Syrian Arab Republic, Paragraph 7, 2013.

[98] Ibid. *supra* note UNGA A/HRC/22/59, Report of the Independent International Commission of Inquiry on the Syrian Arab Republic, Paragraph 7, 2013.

[99] UNGA A/HRC/23/58, Report of the Independent International Commission of Inquiry on the Syrian Arab Republic, Paragraph 8, 2013, https://www.ohchr.org/Documents/HRBodies/HRCouncil/CoISyria/A-HRC-23-58_en.pdf, (accessed 26.09.2019)

See also: UNGA A/HRC/24/46, Report of the Independent International Commission of Inquiry on the Syrian Arab Republic, Paragraph 9, 2013, https://www.ohchr.org/en/Countries/MENARegion/Pages/CoIReports.aspx, (accessed 26.09.2019).

[100] UNGA A/HRC/25/65, Report of the Independent International Commission of Inquiry on the Syrian Arab Republic, Annex VI—Paragraph 1–2, 17, 2014, https://www.ohchr.org/en/Countries/MENARegion/Pages/CoIReports.aspx, (accessed 26.09.2019).

multiples barrel bombs consecutively, ensures maximum damage to intended and unintended targets.[101] When attacking military objectives in densely populated areas with high concentrations of civilians, because of the lack of precision, such areas were treated as a single military objective, which amounts to "area bombardment" prohibited under International humanitarian law (terror among the civilian population).[102] One of the 2014 Report's recommendations to the Syrian Government was to cease using weaponry such as illegal barrel bombs, highlighting that unguided or poorly guided weapons reduces their accuracy and affects civilian areas.[103] In addition, the bombardments with barrel bombs, built as incendiary weapons, violated rules of international humanitarian law, because of their capacity to cause superfluous injury and unnecessary suffering.[104] Barrel bombs have been dropped on civilian-inhabited areas in Idlib, Hama, Al-Raqqa, Aleppo, Dara and Damascus governorates, of which the most intense bombing was in Aleppo city and Daraya in Damascus countryside.[105] The terror from bomb dropping and the substantial civil casualties led to mass displacement.[106]

The 2015 Report mentioned satellite imagery among the procedures collecting evidence and underlined "the complete lack of adherence to the norms of international law".[107] Annex II further documented violations, massacres and other unlawful killing.[108] The first reported use of barrel bombs was in August 2012 in Homs city and in mid 2013 an intensive campaign of barrel bombing of Aleppo city began.[109] The body of evidence collected by the Commission unequivocally demonstrates

[101] UNGA A/HRC/27/60, Report of the Independent International Commission of Inquiry on the Syrian Arab Republic, Paragraph 64, pp. 40, 2014, https://www.ohchr.org/Documents/HRBodies/HRCouncil/CoISyria/A.HRC.27.60_Eng.pdf, (accessed 26.09.2019).

[102] Ibid. *supra* note UNGA A/HRC/25/65, Report of the Independent International Commission of Inquiry on the Syrian Arab Republic, Annex VI—Paragraph 4, 2014, https://www.ohchr.org/en/Countries/MENARegion/Pages/CoIReports.aspx, (accessed 26.09.2019).

[103] Ibid. *supra* note UNGA A/HRC/25/65, Report of the Independent International Commission of Inquiry on the Syrian Arab Republic, Paragraph 158 (a), 2014, https://www.ohchr.org/en/Countries/MENARegion/Pages/CoIReports.aspx, (accessed 26.09.2019).

[104] Ibid. *supra* note UNGA A/HRC/25/65, Report of the Independent International Commission of Inquiry on the Syrian Arab Republic, Paragraph 42, 2014, https://www.ohchr.org/en/Countries/MENARegion/Pages/CoIReports.aspx, (accessed 26.09.2019).

[105] Ibid. *supra* note UNGA A/HRC/25/65, Report of the Independent International Commission of Inquiry on the Syrian Arab Republic, Annex VI—Paragraph 5, https://www.ohchr.org/en/Countries/MENARegion/Pages/CoIReports.aspx, (accessed 26.09.2019).

[106] Ibid. *supra* note UNGA A/HRC/25/65, Report of the Independent International Commission of Inquiry on the Syrian Arab Republic, Paragraph 8, 2014, https://www.ohchr.org/en/Countries/MENARegion/Pages/CoIReports.aspx, (accessed 26.09.2019).

[107] UNGA A/HRC/28/69, Report of the Independent International Commission of Inquiry on the Syrian Arab Republic, Paragraph 1, 2015.

[108] Ibid. *supra* note UNGA A/HRC/28/69, Report of the Independent International Commission of Inquiry on the Syrian Arab Republic, Annex II, pp. 27, 2015.

[109] Ibid. *supra* note UNGA A/HRC/28/69, Report of the Independent International Commission of Inquiry on the Syrian Arab Republic, Paragraph 14, 2015.

"disregard for human life and protections accorded to civilians" and urgently recalls international action needed to address the humanitarian protection needs.[110]

The 2016 Report of the Independent International Commission of Inquiry on the Syrian Arab Republic was presented in the Human Rights Council where it was stated that part of its investigation used satellite imagery in addition to other forms of academic materials.[111] The Report raised the problem of barrel bombs dropped near the Al-Bayan trauma hospital and the adjacent Al-Hakim paediatric hospital in Aleppo city which started a huge fire and killed several persons, including children.[112] It further detailed the consequences of the heavy aerial bombardment including the loss of hospitals and clinics, and loss of humanitarian assistance due to casualties to medical workers, fear and destruction of hospitals.[113] The attacks with barrel bombs contributed to the destruction, including but not limited to destruction of cities, towns and villages, attacks on medical care, attacks on education, attacks on public spaces, cuts to electricity and water, and destruction of cultural heritage.[114]

In 2017, the Commission acknowledged the brutal tactics in the "battle for Aleppo" which resulted in unparalleled suffering for individuals in Syria.[115] In a new report, the Commission further detailed the use of satellite imagery to evidence and confirm violations of International Humanitarian Law, based on the Operational Satellite Application Programme Satellite Imagery Analysis (UNOSAT).

UNOSAT is a technology program delivering imagery analysis and satellite solutions to organizations, including those from the UN system, Member States, IOs and NGOs.[116] Their mission is "to deliver integrated satellite-based solutions for human security, peace and socioeconomic development" to help make a difference in critical

[110] Ibid. *supra* note UNGA A/HRC/28/69, Report of the Independent International Commission of Inquiry on the Syrian Arab Republic, Paragraph 94, 2015.

[111] UNGA A/HRC/33/55, Report of the Independent International Commission of Inquiry on the Syrian Arab Republic, Paragraph 5, 2016, https://undocs.org/A/HRC/33/55, (accessed 26.09.2019).

[112] Ibid. *supra* note UNGA A/HRC/33/55, Report of the Independent International Commission of Inquiry on the Syrian Arab Republic, Paragraph 50, 2016, https://undocs.org/A/HRC/33/55, (accessed 26.09.2019).

[113] Ibid. *supra* note UNGA A/HRC/33/55, Report of the Independent International Commission of Inquiry on the Syrian Arab Republic, Paragraph 50–65, 2016, https://undocs.org/A/HRC/33/55, (accessed 26.09.2019).

[114] UNGA A/HRC/31/68, Report of the Independent International Commission of Inquiry on the Syrian Arab Republic, Paragraph 46, https://www.ohchr.org/Documents/HRBodies/HRCouncil/CoISyria/A-HRC-31-68.pdf, (accessed 26.09.2019).

[115] UN web TV, War crimes committed by all parties in battle for Aleppo—UN mandated inquiry on Syria, 1 March 2017, http://webtv.un.org/live/watch/war-crimes-committed-by-all-parties-in-battle-for-aleppo-—-un-mandated-inquiry-on-syria/5343947021001, (accessed 23.09.2019).

[116] UNITAR, Operational Satellite Applications Programme—UNOSAT, https://unitar.org/unosat/, (accessed 26.09.2019)

UNOSAT is a technology-intensive program delivering imagery analysis and satellite solutions to relief and development organizations within and outside the UN system to help make a difference in critical areas such as humanitarian relief, human security, strategic territorial and development planning. UNOSAT develops applied research solutions keeping in sight the needs of the beneficiaries at the end of the process. UNOSAT is UNITAR's Operational Satellite Applications Programme.

areas. This activity is in keeping with the mandate given to UNITAR. Integrating field collected data with remote sensing imagery and Geographic Information Systems (GIS), UNOSAT produces concrete and usable applications including for human rights.[117] Satellite imagery over Syria was used to confirm a crater consistent with an aerial bomb.[118] Also, in proving the claim made by victims that the attack was the result of an air bombardment, the investigation included analysis of satellite imagery "showing impact consistent with the use of air-delivered munitions".[119] Satellite imagery was used with early warning reports, witness testimony, forensic evidence gathered at the site and data provided by Member States. Such evidence proves the use of air-delivered munitions and implicates Syrian forces in the attacks.[120]

The satellite-based imagery damage assessment in Aleppo Governorate conducted by UNOSAT is an example of the usefulness of satellite-based technology in the investigation. The effort to identify where the Red Crescent/UN aid convoy suffered an alleged attack was complemented by satellite images. Using satellite imagery acquired on different dates, UNOSAT identified where the convoy was visibly damaged and scorched. Additional imagery showed the area cleaned up with trucks removed.[121] Classic evidence would not have been useful if not witnessed in real time which is why the images from satellites provided a comparative image in different periods, representing a vital body of evidence for the investigation. Other examples of satellite imagery in the investigation were the damage assessment for the alleged Aleppo's hospitals attacks, including evidence of new debris near the hospital structure, a damaged corner of the hospital structure and debris on the street, destroyed walls and damaged facades.[122] The latter included analysis of satellite images looking for evidence about the alleged nine barrel bombs dropped on 16 November, 2016 on Al-Hakim, Al Zahraa and Al-Bayan hospitals and the central blood bank.[123]

In the 2018 Report, as part of the methodology for its investigation, the Commission continued to collect, review and analyse satellite imagery in addition to

---

117 UNITAR, Operational Satellite Applications Programme—UNOSAT, https://unitar.org/unosat/what-we-do, (accessed 23.09.2019).

118 UNGA A/HRC/34/64, Report of the Independent International Commission of Inquiry on the Syrian Arab Republic, Paragraph 85, 2017, https://undocs.org/A/HRC/34/64, (accessed 26.09.2019).

119 Ibid. *supra* note UNGA A/HRC/34/64, Report of the Independent International Commission of Inquiry on the Syrian Arab Republic, Paragraph 85, 2017, https://undocs.org/A/HRC/34/64, (accessed 26.09.2019).

120 Ibid. *supra* note UNGA A/HRC/34/64, Report of the Independent International Commission of Inquiry on the Syrian Arab Republic, Paragraph 87, 2017, https://undocs.org/A/HRC/34/64, (accessed 26.09.2019).

121 UNITAR/ UNOSAT, UN/SARC Aid Convoy Damage Assessment Urum al Kubrah Aleppo Governorate, Syria, 19 December 2016, https://www.ohchr.org/Documents/HRBodies/HRCouncil/CoISyria/UNOSAT_Satellite_Imagery_1.pdf, (accessed 23.09.2019).

122 UNITAR/UNOSAT, Alleged Aleppo's hospitals attacks, Damage Assessment, 30 September 2016, https://www.ohchr.org/Documents/HRBodies/HRCouncil/CoISyria/UNOSAT_Satellite_Imagery_2.pdf, (accessed 23.09.2019).

123 Ibid. *supra* note Satellite Imagery 2, pp. 2.

photographs, videos and medical records together with interviews conducted in the region and from Geneva. Communications from governments and NGOs were used, as were UN reports.[124] The documentary and material evidence analysed by the Commission for two attacks on the hospital in Afrin city in 16 March, 2018 evidenced the damage caused to the hospital and was provided by satellite imagery published on the website of the Commission of Inquiry.[125]

The latest report in 2019 of the Independent International Commission of Inquiry on the Syrian Arab Republic continued to mention that as part of its investigation, the Commission had collected, reviewed and analysed various materials including satellite imagery in addition to photographs, videos and medical records, communications from governments and NGOs.[126] Information in relation to the foregoing incidents analysed raised "serious concerns that war crimes and other violations of international law may have been committed".[127]

The capacity of the domestic judicial system is questionable in addressing violations of human rights and international humanitarian law. Various challenges refer to the prosecution of criminals and the implementation of reforms. As was pointed out in the Commission's Report in 2012, in accordance with international law, the responsibility to investigate, prosecute and punish international crimes and other gross violations rests first and foremost with the Syrian state.[128] The crimes against humanity and other gross violations documented in the commission's reports have been committed, however, within a system of impunity. At that specific time, the need for "profound structural reforms in the political, justice and security sectors (…) to break the culture of impunity and to deliver justice to the victims" was highlighted. However, there were no notable differences. The Commission took into consideration in 2019 that administrative and executive structures in Syria were described by the interviewees as largely ineffective and incapable of addressing accusations as to the unlawful conduct of dozens of armed groups.[129]

---

[124] UNGA A/HRC/39/65, Report of the Independent International Commission of Inquiry on the Syrian Arab Republic, Paragraph 23, 2018, https://undocs.org/A/HRC/39/65, (accessed 26.09.2019).

[125] UNGA A/HRC/39/65, Report of the Independent International Commission of Inquiry on the Syrian Arab Republic, Paragraph 23, 2018, https://undocs.org/A/HRC/39/65, (accessed 26.09.2019).

[126] UNGA A/HRC/40/70, Report of the Independent International Commission of Inquiry on the Syrian Arab Republic, 2019, https://www.ohchr.org/Documents/HRBodies/HRCouncil/CoISyria/A_HRC_40_70.pdf, (accessed 23.09.2019).

[127] Ibid. *supra* note UNGA A/HRC/40/70, Report of the Independent International Commission of Inquiry on the Syrian Arab Republic, Paragraph 48, 2019, https://www.ohchr.org/Documents/HRBodies/HRCouncil/CoISyria/A_HRC_40_70.pdf, (accessed 23.09.2019).

[128] Ibid. *supra* note A/HRC/19/69, Report of the Independent International Commission of Inquiry on the Syrian Arab Republic, Paragraph 127, pp.22, 2012, https://undocs.org/en/A/HRC/19/69, (accessed 26.09.2019).

[129] Ibid. *supra* note UNGA A/HRC/40/70, Report of the Independent International Commission of Inquiry on the Syrian Arab Republic, Paragraph 7, pp. 14, 2019, https://www.ohchr.org/Documents/HRBodies/HRCouncil/CoISyria/A_HRC_40_70.pdf, (accessed 23.09.2019).

For this reason, among its various recommendations, in 2019 the Commission emphasized the need for legislation, legal and administrative mechanisms to process civil acts, and property registration that comply with international human rights law. Satellite imagery could further support the protection of human rights as it proved to do at the international level, in particular when using satellite imagery in the Commission investigation, and is a means to provide evidence to courts in domestic law.

2.4.4 Monitoring Refugees in Crisis Areas

The 1951 Convention on the Status of Refugees defines a refugee as person who: "owing to well-founded fear of being persecuted for reasons of race, religion, nationality, membership of a particular social group or political opinion, is outside the country of his nationality and is unable, or owing to such fear, is unwilling to avail himself of the protection of that country; or who, not having a nationality and being outside the country of his former habitual residence as a result of such events, is unable or, owing to such fear, is unwilling to return to it".[130]

Serious violations of human rights were committed by the Syrian state or non-state actors over a territory and many examples were provided in the Reports of the Commission of Inquiry. From the investigation and in relation to Human Rights Violations in the Syrian Arab Republic it can be reasonably concluded that the consequences of such violations have led to increased forced displacement over the years.

An increase can be noticed in the global trends of forced displacement for reasons such as disasters or conflict. Total registered Syrian refugees, last updated on 9 May 2019, was 5,626,497 people of which 3,606,208 were in Turkey; 938,531 in Lebanon; 660,393 in Jordan, 253,371 in Iraq, 132,281 in Egypt and 35,713 in other states in North Africa.[131] Globally, 44,400 new displacements happen every day. There are 68.5 million forcibly displaced worldwide, of which 25.4 million are refugees and 19.9 million are under UNHCR's mandate; 40.0 million displaced people and 3.1 million asylum-seekers.[132] The highest forced displacement associated with conflict and violence in 2018 was registered in Ethiopia, the Democratic Republic of the Congo (DRC) and Syria.[133] In Syria, in 2018 there were 1,649,000 displacements, which does not include the huge number of Syrians displaced internally within the

[130]UN General Assembly, *Convention Relating to the Status of Refugees*, 28 July 1951, United Nations, Treaty Series, vol. 189, p. 137; Article 1, (A) (2), https://www.unhcr.org/3b66c2aa10, (accessed 18.05.2019).

[131]Source UNHCR, https://data2.unhcr.org/en/situations/syria, (accessed 18.05.2019).

[132]UNHCR—UN Refugee Agency, Global trends: Forced Displacement in 2017, https://www.unhcr.org/5b27be547.pdf, (accessed 18.05.2019).

[133]IDMC—Internal displacement monitoring centre, Norwegian Refugee Council. GRID 2019, Global Report on Internal Displacement, (2019) Figure 5. http://www.internal-displacement.org/sites/default/files/publications/documents/2019-IDMC-GRID.pdf, (accessed 19.05.2019).

country, due to conflict. The population of concern in Syria includes 6.2 million people that are internally displaced.[134]

Forced displacement may rise everywhere when human rights violations increase significantly, and, in particular, one of the problems in Syria as referred to above was the weak rule of law. Domestic legislation and government policies and practices have failed to prevent or halt violations of due process and human rights, which in fact have resulted in amplifying the risk that violations will occur. Another element that leads to forced displacement is the impunity of individuals and authorities that violate the rights of individuals or groups. The system has not been effective in punishing the individuals held responsible for those violations, meaning that the victims or their families did not receive redress or reparation. Impunity increases the likelihood that violations will continue in the future.

Also, political instability can make a state more prone to take actions that are detrimental to the society as a whole and which in crisis areas will ultimately violate human rights and lead to discrimination. Discrimination in this context may refer to exclusion or restriction of individuals with different political beliefs. Identifying the above-mentioned types of hazards, including human rights violation, discrimination, weak rule of law, impunity and political instability is part of the refugee emergency risk analysis.[135]

Demographic and location data are "a robust evidence base for effective and efficient policy responses and programmatic interventions" that help the responsible organizations to address the needs of vulnerable groups and help ensure that "no one is left behind".[136] One of the UNHCR concerns is to address monitoring of refugees, which is fundamental for assuring the right policy. Estimating the size of forcibly displaced populations is key to documenting their plight and allocating sufficient resources to their assistance. However, the analysis is very complex and is often not done for various reasons and, in particular, during the acute phase of displacement, such as due to methodological challenges and inaccessibility of data. For this reason, there is a huge potential for the use of Very High Resolution (VHR) satellite imagery to assist the UNHCR to remotely estimate forcibly displaced populations.[137]

Internally displaced people are forced to flee or leave their homes but without crossing an internationally recognized border.[138] The analysis undertaken by the Internal Displacement Monitoring Centre shows that significant data gaps exists in

---

[134]UNHCR, Syria January–March 2019 operational update, 2019, https://www.unhcr.org/sy/wp-content/uploads/sites/3/2019/05/Operational-update-2019-Q1.pdf, (accessed 18.05.2019).

[135]UNHCR, Risk analysis and monitoring—refugee emergencies, https://emergency.unhcr.org/entry/252070/risk-analysis-and-monitoring-refugee-emergencies, (accessed 18.05.2019).

[136]Ibid. *supra* note. UNHCR—UN Refugee Agency, Global trends: Forced Displacement in 2017, Chapter 8 Demographic and Location Data pp. 57, https://www.unhcr.org/5b27be547.pdf, (accessed 18.05.2019).

[137]Checchi, F., Stewart, B.T., Palmer, J.J., Grundy, C., Validity and feasibility of a satellite imagery-based method for rapid estimation of displaced populations, 23 January 2013, https://www.ncbi.nlm.nih.gov/pmc/articles/PMC3558435/, (accessed 23.09.2019).

[138]OHCHR, Questions and Answers about Internally Displaced Persons, https://www.ohchr.org/EN/Issues/IDPersons/Pages/Issues.aspx, (accessed 18.05.2019).

the investigation and the negative consequence is the lack of information and limited understanding of the number of people displaced.[139] In addition, lack of data means slow progress in managing and reducing urban displacement.

Conflict and violence has negatively impacted internal displacement in the Middle East and North Africa, in particular Yemen, Iraq, Iran, as well as Syria and Libya. Addressing the conflict in Syria, the Global Report on Internal Displacement (GRID) 2019 supported the idea that the conflict is apparently nearing its end.[140] Despite the decline in conflict in 2018, Syria continued to trigger one of the largest population movements in the world, with the highest number in the region. When referring to reconstruction of major urban centres such as Aleppo and Raqqa, the importance of the displaced individuals to be able to return in safety and dignity was underlined. Due to the large number of displaced people, nearly half of the country's population, the return of refugees and internally displaced people is a huge challenge. Monitoring is part of supporting the government to address its displacement crisis with comprehensive policies that would prevent a conflict from developing again.

Libya is also among the areas with urban displacement driven by escalating conflict and insecurity.[141] The deteriorating economic situation in Tripoli is fuelling the ongoing conflict and as a result, almost 33,000 people have been displaced. Libya is known as the point of departure for people attempting to cross the Mediterranean Sea to Europe but also as "a final destination for economic migrants from West Africa". Assuring security has been described as a "daunting task with relatively little promise" mainly because of the vastness of the territory.[142] Satellite-based monitoring could be a solution to monitor and provide the necessary data and location.

In Iraq, despite the official ending of the conflict on 9 December 2017, almost two million people were recorded at the end of 2018 to be living in displacement, mainly because of the lack of livelihood opportunities, the presence of landmines and unexploded ordinance, and flood and drought.[143] The challenge has been to gather information about the people displaced or returning but such information has not been available to determine the situation of all.

Many areas are not accessible, such as the North Sinai Governorate in Egypt with more than 15,000 displacements, where the statistics and numbers provided could only be "an underestimate given that the area is inaccessible, and the figure was

[139] Ibid. *supra* note IDMC—Internal displacement monitoring centre, Norwegian Refugee Council. GRID 2019, Global Report on Internal Displacement, pp. iv, 2019.

[140] IDMC—Internal displacement monitoring centre, Norwegian Refugee Council. GRID 2019, Global Report on Internal Displacement, pp. 25, 2019, http://www.internal-displacement.org/sites/default/files/publications/documents/2019-IDMC-GRID.pdf, (accessed 19.05.2019).

[141] IDMC—Internal displacement monitoring centre, Norwegian Refugee Council. GRID 2019, Global Report on Internal Displacement, pp. 26, 2019.

[142] Ibid. *supra* note IDMC—Internal displacement monitoring centre, Norwegian Refugee Council. GRID 2019, Global Report on Internal Displacement, (2019), pp. 27.

[143] Ibid. *supra* note IDMC—Internal displacement monitoring centre, Norwegian Refugee Council. GRID 2019, Global Report on Internal Displacement, (2019), pp. 21.

compiled using satellite imagery and eyewitness accounts".[144] The IDMC's estimate of the total number of internal displaced persons in Egypt was mainly based on the Human Rights Watch reports from 2015 and 2018 that relied on a combination of satellite imagery analysis and interviews with affected families.[145]

Humanitarian responses to solve the internal displacement crisis should include national responsibility and leadership for the monitoring and protection of the human rights of refugees and/or internally displaced persons. Interoperable data are necessary to support local action, particularly satellite-based data related to respecting human rights as well as monitoring of refugee and Internally displaced persons camps using SENTINEL-2 Imagery.[146]

Since 2001, the technological assistance of satellite images SPOT or LANDSAT and GNSS have been acknowledged by UNHCR in continuous monitoring of the refugee and refugee-affected regions mainly for environmental aspects around camps. In one particular case, data was gathered on the origin of refugees, as well as the number and location of different ethnic groups within the camp, which was useful for camp planning and management.[147]

The European Association of Remote Sensing Companies (EARSC) in an article in 2015 underlined that satellite EO technologies can and will play an increasingly important role in monitoring the movement of migrants and refugees.[148] The need for reliable information was directly linked to the role and capabilities of the satellites. It was stated that a responsive solution is needed to monitor and control the movement of migrants across the Mediterranean Sea and through eastern European borders. Information on migrant movements is limited and the particularity of satellite technology is that it can efficiently cover a large geographic area of land and water.

Three factors were identified as crucial in detection and identification of migrants: (i) the capacity to monitor activities over large areas of land and sea; (ii) the ability to collect detailed data to take proportionate and measured actions; (iii) the capability to disseminate up-to date observations to decision makers.[149] Synthetic Aperture Radar (SAR) complemented by VHR optical images is a solution for monitoring and detecting activities over narrow areas. In addition, UAVs connected and directed by satellite can be used to supply very high-resolution real-time surveillance. The

144 Ibid. *supra* note IDMC—Internal displacement monitoring centre, Norwegian Refugee Council. GRID 2019, Global Report on Internal Displacement, (2019), pp. 22.

145 Ibid. *supra* note IDMC—Internal displacement monitoring centre, Norwegian Refugee Council. GRID 2019, Global Report on Internal Displacement, (2019), pp. 129.

146 Wendt, L., Lang, S., Rogenhofer, E., Monitoring of Refugee and Camps for Internally Displaced Persons Using Sentinel-2 Imagery—A Feasibility Study, January 2017, https://www.researchgate.net/publication/318075028_Monitoring_of_Refugee_and_Camps_for_Internally_Displaced_Persons_Using_Sentinel-2_Imagery_-_A_Feasibility_Study, (accessed 23.09.2019).

147 UNHCR, Monitoring and evaluation, 01 January 2001, https://www.unhcr.org/protection/environment/3b03d9a94/monitoring-evaluation.html, (accessed 18.05.2019).

148 Wheeler, J., People Watching: Tracking and Monitoring Migration, http://earsc.org/news/people-watching-tracking-and-monitoring-migration, (accessed 18.05.2019).

149 Ibid. *supra* note Wheeler, J., People Watching: Tracking and Monitoring Migration.

imminent deployment of large constellations of satellites in LEO could also offer the satellite capability to disseminate information on very large areas in real time with seamless connectivity.

2.4.5 Attacks on UN Camps in Africa

In 1987, the UNHCR Executive Committee at its 38th Session discussed the Resolution 48 (XXXVIII) about the Military or Armed Attacks on Refugee Camps and Settlements. The Resolution 48 (XXXVIII) highlighted that refugee camps and settlements have an exclusively civilian and humanitarian character and reminded that unlawful attacks on refugees and asylum seekers in different areas of the world was strongly condemned by the General Assembly Resolution 39/140 (1984) referring to all violations of the rights and safety of refugees and asylum-seekers. The UNHCR urged States to abstain from military or armed attacks on refugee camps and settlements and provided a number of guidelines to enhance the protection of refugee camps and settlements. The Executive Committee of the UNHCR stated that States have a duty to co-operate with the High Commissioner in the performance of his humanitarian protection and assistance functions.[150] In 2011, satellite images showed refugee camp attacks.[151] People were killed in an attack on a Catholic mission with 20,000 refugees in the Central African Republic town of Alindao, around 300 km of the capital Bangui. Such an attack reflects the irresponsibility and contempt for civilians, including the disproportionate and indiscriminate use of force, said the UN MINUSCA.[152] The attacks against the peacekeepers may constitute a war crime.

The UN Camp in Mali for the Multidimensional Integrated Stabilization Mission (MINUSMA) was attacked in January 2019 by unidentified assailants and resulted in more than 25 injured and ten dead UN peacekeepers. The UN Security Council stated that targeting UN peacekeepers "constitutes a basis for sanctions" in particular as a war crime. Other attacks against UN peacekeepers occurred in Mali in August 2017 (attack on MINUSMA headquarters in Timbuktu) and in November 2017.[153]

[150] Executive Committee 38th Session, Military or Armed Attacks on Refugee Camps and Settlements No. 48 (XXXVIII)—1987, https://www.unhcr.org/excom/exconc/3ae68c951c/military-armed-attacks-refugee-camps-settlements.html, (accessed 8.08.2019).

[151] Zapata, M., Escalating Crisis in Sudan: Satellite Imagery Shows Increased SAF Air Capacity, Refugee Camp Attacks, 14 November 2011, https://enoughproject.org/blog/escalating-crisis-sudan-refugee-camp-attacks-and-satellite-imagery-showing-increased-saf-air, (accessed 23.09.2019).

[152] Stubley, P., At least 42 killed in attack on refugee base in Central African Republic, 17 November 2018, https://www.independent.co.uk/news/world/africa/attack-refugee-camp-militia-central-african-republic-alindao-seleka-anti-balaka-a8639206.html, (accessed 8.08.2019).

[153] Andone, D., Ten UN peacekeepers killed in attack in Mali, 21 January 2019, https://edition.cnn.com/2019/01/20/africa/un-peacekeepers-killed-mali/index.html, (accessed 8.08.2019).

Secretary General Antonio Guterres condemned the armed attacks and reaffirmed his support for the Malian authorities.[154] AI has presented and asked the UN for an investigation. Another camp in Alindao for internally displaced persons (IDPs) was also attacked killing between 70 and 100 people. In fact, AI stated that the camp was under the protection of UN MINUSCA when it was attacked by rebels. It was stated that the UN force did not fire back and the perpetrators escaped unharmed.[155]

PLANET IMAGES show white tents and containers, and capture the expansion of Syrian displaced people along the Turkish border.[156] The UNHCR base DusitD2 complex hosting around 230,000 refugees in Kenya was attacked in 2019.[157] Another attack on UN camps happened in Libya, which killed more than 44 migrants at a detention centre.

2.4.6 *Illicit Arm Traffic*

The UN Office on Drugs and Crime (UNODC) has acknowledged the weak preventive, regulatory and security measures in place to control weapons and their movements. The difficulties include prevention, detection, investigation and prosecution of these crimes. Strengthening of the criminal justice response implies the use of special investigative techniques.[158] SDG 16, to "Promote peaceful and inclusive for sustainable development, provide access to justice for all and build effective, accountable and inclusive institutions at all levels" targets a significant reduction in illicit financial and arms flows, and the strengthening of the recovery and return of stolen assets, as well as combating all forms of organized crime by 2030.

Diverse trafficking methods and routes make limitless the number of small arms smuggled. Poorly or corrupted land borders facilitate the illicit transfer of small arms, on foot or by truck. Some of this trafficking is large-scale and systematic. Arms embargo violations in Somalia documented by UN experts have revealed the

---

[154] Dicko, D., UN News, Guterres condemns armed attack against UN peacekeepers in Mali, 20 May 2019, https://news.un.org/en/story/2019/05/1038761, (accessed 8.08.2019).

[155] RFI, Amnesty slams UN mission in CAR over attack on IDP camp, 20 December 2018, http://en.rfi.fr/africa/20181218-amnesty-UN-minusca-investigation-attack-CAR-idp-camp-seleka-balaka, (accessed 08.08.2019).

[156] Banco, E., Latest Satellite Imagery Captures Expansion of Syrian IDP Camps on Turkish Border, 2 September 2016, https://medium.com/planet-stories/latests-satellite-imagery-captures-expansion-of-syrian-idp-camps-on-turkish-border-383634ecfdf5, (accessed 08.08.2019).

[157] Fryberg, M., Kenya terror attack suspects traced to Dadaab refugee camp, 4 February 2019, https://www.iol.co.za/news/africa/kenya-terror-attack-suspects-traced-to-dadaab-refugee-camp-19120342, (accessed 08.08.2019).

[158] Grassi, S., United Nations Office on Drugs and Crime, Countering Illicit Arms Trafficking and its Links to Terrorism and Other Serious Crime. UNODC's Global Firearms Programme, 17 May 2017, https://www.un.org/sc/ctc/wp-content/uploads/2017/05/Simonetta-UNODC-at-CTED_May2017v2.pdf, (accessed 24.05.2019).

delivery of guns to Somali militias by Ethiopian truck convoys.[159] Africa's rivers and coasts are also used by traffickers to smuggle small arms. For example, waterproofs sacks are attached to the bottom of boats. Also, aircraft are used to ferry weapons inter-continentally and regionally,[160] for example the shipment of Ukrainian small arms ammunition to the Ivory Coast in July 2000. The availability of guns from the black market has led to a "gun culture" in many countries in Africa.

International arms flows monitoring through satellite images could be included as part of the new innovative tools for assessing arms flows. Limitations include the significant resources needed for the large-scale collection and processing of such data.[161] More effective data collection efforts require careful consideration of the new tools. Too costly for others, intelligence agencies monitor arms flows that involve their specific national interests, including large-scale and high-resolution satellite or aerial images.[162] Satellite imagery has become cheaper and thus more accessible. NGOs detect the deployment of major weapons and military formations. "When investigating arms flows satellite images can corroborate other information or they can be a lead for further investigations". While satellite images have limited usefulness for small arms traffic individually, they can detect convoys based on information provided on the ground. Satellite images by themselves will not reveal the supplier of the weapons but will help to make the conclusions and the assessment. Satellite imagery may be able to confirm a minimum number of weapons but not the exact number, which is a limitation and comes with relatively high costs.[163]

2.4.7 *Monitoring Group on Somalia and Eritrea UN Security Council*

The UN Security Council, at its meeting in December 2003, conferred a mandate to the Monitoring Group focusing on violations of the arms embargo, including transfers of ammunition, single-use weapons and small arms.[164] The Monitoring Group

[159] Schroeder, M., Lamb, G. The Illicit Arms Trade in Africa: A Global Enterprise. In African analyst, pp. 71, July 2018, https://www.researchgate.net/publication/326462123_The_Illicit_Arms_Trade_in_Africa_A_Global_Enterprise, (accessed 24.05.2019).

[160] Ibid. *supra* note Schroeder, M., Lamb, G. The Illicit Arms Trade in Africa: A Global Enterprise. In African analyst, pp. 71, July 2018.

[161] Wezeman, P., International arms flows: monitoring, sources and obstacles. Clingendael Report, January 2018, https://www.clingendael.org/sites/default/files/2018-01/Report_International_Arms_Flows.pdf, (accessed 24.05.2019).

[162] Ibid. *supra* note Wezeman, P., International arms flows: monitoring, sources and obstacles. Clingendael Report, January 2018, pp. 7.

[163] Ibid. *supra* note Wezeman, P., International arms flows: monitoring, sources and obstacles. Clingendael Report, January 2018, pp. 14.

[164] UN Security Council, S/RES/1519 2003), Resolution 1519 (2003) Adopted by the Security Council at its 4885th meeting, Paragraph 2, pp. 2, 16 December 2003, http://unscr.com/en/resolutions/doc/1519, (accessed 26.09.2019).

on Somalia started reporting in 2004 taking into consideration the findings of the two previous panels of experts that started their investigation on Somalia in 2003 following the first report of the team of experts in 2002.[165] Since its first report, the Monitoring Group has evidenced the challenges and difficulties in obtaining information in the field and has adapted its standard of evidence to the Somali context.[166] Analysing satellite imagery was necessary for gathering evidence and was acknowledged in the Reports.[167]

In its first report in 2004, the relationship between the security situation in Somalia and the violations of the arms embargo illustrated "an apparent never-ending cycle of violent and deadly confrontations" involving "the use of a wide variety of arms by an equally wide variety of individuals and groups".[168] The arms traffic was investigated based on persistent reports of arms coming into Somalia from the west across the

---

(a) To investigate the violations of the arms embargo covering access to Somalia by land, air and sea;

(b) To detail information and make specific recommendations in relevant areas of expertise related to violations and measures to give effect to and strengthen the implementation of the arms embargo in its various aspects;

(c) To carry out field-based investigations in Somalia, where possible, and in States neighbouring Somalia and other States, as appropriate;

(d) To assess the progress made by the States in the region to implement fully the arms embargo, including through a review of national customs and border control regimes;

(e) To provide the Committee in its final report with a draft list of those who continue to violate the arms embargo inside and outside Somalia, and their active supporters, for possible future measures by the Council;

(f) To make recommendations based on its investigations and the previous reports of the Panel of Experts (S/2003/223 and S/2003/1035) appointed pursuant to resolutions 1425 (2002) and 1474 (2003).

165 UN Security Council S/2004/604, (2004) "Report or the Monitoring Group on Somalia submitted in accordance with resolution 1519 (2003)", Paragraph 24.

166 Ibid. *supra* note UN Security Council S/2004/604, (2004) "Report or the Monitoring Group on Somalia submitted in accordance with resolution 1519 (2003)", Paragraph 36:

(a) Collection of information on events and topics from multiple sources where possible;

(b) Collection of information from sources with first-hand or quasi-first-hand knowledge of events;

(c) Looking for the consistency of the patterns of information and comparing them with new information;

(d) Looking at the totality of the information obtained on issues, topics and events and developing a mosaic of the information. Determining how the new information compares with the substance, tone and feel of the information already collected;

(e) Continuously factoring in the expertise and judgement of the relevant expert of the Group and the collective assessment of the Group;

(f) Continuously seeking out documentary support of the information collected.

167 United Nations Security Council, Reports Panel of Experts, Somalia Sanctions Committee, https://www.un.org/securitycouncil/sanctions/751/work-and-mandate/reports, (accessed 23.09.2019).

168 Ibid. *supra* note UN Security Council S/2004/604, Report or the Monitoring Group on Somalia submitted in accordance with resolution 1519 (2003), Paragraph 64, 2004.

Ethiopian border.[169] In addition, the Monitoring Group reported a clandestine flow of arms in Kenya from Somalia.[170]

In 2005, the Monitoring Group revealed that during September 2004 to February 2005 arms flows continued at an alarming rate and in large quantities, with organized parties engaged in the arms transactions such as traders, smugglers and transnational criminal groups, with shipment originating in Europe, Asia and the Middle East.[171] Pertinent information for the investigation of arms embargo violations followed the chain of events for each violation, including the buyers, the sellers, the middlemen, the shipment and shipping companies involved, other modes of transport, the means and methods of payment.[172] Arms were mainly transported by sea due to the relative ease to arrange transport and cost effectiveness, relying less on air transport.[173] The traditional border crossing points also constituted an important element in the arms embargo violations.

Referring to evidence, in the 2005 Report, the Monitoring Group referred to the standard of verification mentioned back in 2004. The difficulties and challenges of gathering evidence necessitated the development of the concept of source circles and the case study model.[174] When describing the pirates off the coast of Somalia, the second Report in 2005 evidenced them as trained fighters using speedboats equipped with satellite technology, in particular satellite phones and GNSS, helping them in receiving instructions from land-based headquarters for seizing ships and obtaining money through ransom demands.[175] Recalling the arms trade, the network of smugglers relies heavily on the port of Bossaso and the road infrastructure of Puntland.[176]

---

[169] Ibid. *supra* note UN Security Council S/2004/604, Report or the Monitoring Group on Somalia submitted in accordance with resolution 1519 (2003), Paragraph 65, 2004.

[170] Ibid. *supra* note UN Security Council S/2004/604, Report or the Monitoring Group on Somalia submitted in accordance with resolution 1519 (2003), Paragraph 66, 2004.

[171] UN Security Council S/2005/153, Report of the Monitoring Group on Somalia submitted in accordance with resolution 1558 (2004), Paragraphs 9–10, 2005.

[172] UN Security Council S/2005/153, Report of the Monitoring Group on Somalia submitted in accordance with resolution 1558 (2004), Paragraph 6, 2005.

[173] UN Security Council S/2005/153, Report of the Monitoring Group on Somalia submitted in accordance with resolution 1558 (2004), Paragraphs 12-14, 2005.

[174] UN Security Council S/2005/153, Report of the Monitoring Group on Somalia submitted in accordance with resolution 1558 (2004), Annex 1, Paragraph 3 (a) (b), 2005:

(a) ***Source circles.*** *Deliberate and systematic effort to gain access to those involved in arms embargo violations by way of individuals who have direct knowledge or know people who have direct knowledge of details of violations;*

(b) ***Case study model.*** *The case study model is specific arms shipment/ transaction based. It is a tool that helped the Monitoring Group to systemize and standardize its investigative work and includes optimum questions that attempt to answer essential elements of information for all shipments investigated by the Monitoring Group.*

[175] UN Security Council S/2005/625, Report of the Monitoring Group on Somalia submitted in accordance with resolution 1587 (2005), Paragraph 100–105, 2005.

[176] Ibid. *supra* note UN Security Council S/2005/625, Report of the Monitoring Group on Somalia submitted in accordance with resolution 1587 (2005), Paragraph 14, 2005.

In 2006, arms flows continued to be reported by the Monitoring Group, with evidence being also gathered from interviewing government officials in the region, representatives of diplomatic missions, civil society organizations, "knowledgeable individuals" from Somali civil society and opposition members. The report said that ideological vested interests had developed during the mandate of the Monitoring Group, with the objective of creating an Islamic State, run by Sharia law and having as consequence the continuation of arms embargo violations.[177] In the second Report of 2016, the Monitoring Group advanced severe conclusions about the situation in Somalia, such as no implementation of past recommendations, the security situation continuing to deteriorate, resulting in an explosion of arms flows, wider militarization of society and contenders for control, all in violation of the arms embargo.[178] Among the Monitoring Group recommendations, was included the request to implement total border surveillance, involving a combination of sea, air and land military forces supported by other organizations.[179]

The Report of 2007 of the Monitoring Group stated that Ethiopia had introduced its own arms and troops in Somalia, Eritrea being the main clandestine source and conduit for arms supplies.[180] Arms appear to have been bought in Somalia via clandestine routes and intended for use by key Somali actors. The difficulties in monitoring developments continued to be reported by the Monitoring Group together with the worrying situation of arms trafficking, "Somalia is literally awash with arms". Included in the recommendations was the priority of locating and dealing with hidden arms caches.[181]

In 2008, the Monitoring Group started by describing the embargo as having had limited impact on the conflict in Somalia, because the parties had the capacity to receive arms, further mentioning the border patrols were not efficient in controlling borders in violation of the arms embargo "Cross-border arms shipments in violation of the arms embargo either reach the different parties to the conflict directly, or fuel the seven different arms markets (…)".[182] In addition to the previous information concerning members of the Transitional Federal Government buying arms at the market in Mogadishu, the Monitoring Group revealed that information had been received on sales of arms by "prominent officials of the security sectors of the Government,

---

177 UN Security Council S/2006/229, Report of the Monitoring Group on Somalia submitted in accordance with resolution 1630 (2005), Paragraph 137–138, 2006.

178 UN Security Council S/2006/913, Report of the Monitoring Group on Somalia submitted in accordance with resolution 1630 (2005), Paragraphs 235–236, 2006.

179 Ibid. *supra* note UN Security Council S/2006/913, Report of the Monitoring Group on Somalia submitted in accordance with resolution 1630 (2005), Paragraph 238, 2006.

180 UN Security Council S/2007/436, Report of the Monitoring Group on Somalia submitted in accordance with resolution 1724 (2006), Paragraph 10, 2007, https://www.undocs.org/S/2007/436, (accessed 26.05.2019).

181 Ibid. *supra* note UN Security Council S/2007/436, Report of the Monitoring Group on Somalia submitted in accordance with resolution 1724 (2006), Paragraph 120 (c), 2007.

182 UN Security Council S/2008/274, Report of the Monitoring Group on Somalia submitted in accordance with resolution 1766 (2006), pp. 6, 2008, https://www.undocs.org/S/2008/274, (accessed 26.05.2019).

Ethiopian officers and Ugandan officers of the AU Mission in Somalia".[183] For the methodology of investigation, the practice of due process was added in addition to the evidentiary standards and verification processes outlined in previous reports. The practice of due process included letters sent to states, with only ten replying from 38 states notified.[184] Among the non-replying states was also a lack of cooperation from the AU Commission.

In its second report of 2008, the Monitoring Group stated that imports from Yemen remained the most consistent source of arms, ammunition and military materiel to Somalia, feeding the armed opposition and criminal groups. Also, a different source of arms, ammunition and military materiel to Somalia was the sale of weapons by Transitional Federal Government security officials, who are challenged by corruption and loss of authority. In particular, external support to the forces of the Transitional Federal Government are mostly non-authorized and constitute violations of the arms embargo, being diverted to private purposes after the declaration of ammunition as being "expended during combat" which in fact was sold for money.[185] As part of the methodology of investigation, in the second Report in 2018 the systematic effort to gain access to individuals involved in arms embargo violations was described.

The next report of the Monitoring Group was in 2010, when it was underlined that efforts to restore peace and security to Somalia were being critically undermined by a war economy which resulted in corruption of the state institutions at all levels.[186] As a result of domestic division and a primary symptom of the war economy was the growth of piracy. Referring to the methodology for gathering evidence, it was highlighted that in the absence of judicial recourse, the Monitoring Group would consider as "beyond a reasonable doubt" the information obtained from at least three credible and independent primary sources, in addition to in situ expert observations.[187] Information offered to the Monitoring Group revealed that arms continued to be sent to Somalia from Eritrea and Yemen.[188] In particular, Yemen remained "Somalia's principal commercial market for weapons" while Ethiopia was a principal state supplier of arms, ammunition and training. Finally, the transactions from Eritrea focused on supporting armed opposition groups.[189]

[183] Ibid. *supra* note UN Security Council S/2008/274, Report of the Monitoring Group on Somalia submitted in accordance with resolution 1766 (2006), pp. 6, 2008.

[184] Ibid. *supra* note UN Security Council S/2008/274, Report of the Monitoring Group on Somalia submitted in accordance with resolution 1766 (2006), pp. 44, Paragraphs 217–222, 2008.

[185] UN Security Council S/2008/274, Report of the Monitoring Group on Somalia submitted in accordance with resolution 1766 (2006), pp. 6, 2008.

[186] UN Security Council S/2010/91, Report of the Monitoring Group on Somalia submitted in accordance with resolution 1853 (2008), pp. 6–7, 2010, https://www.undocs.org/S/2010/91, (accessed 26.05.2019).

[187] Ibid. *supra* note UN Security Council S/2010/91, Report of the Monitoring Group on Somalia submitted in accordance with resolution 1853 (2008), Paragraph 10, 2010.

[188] Ibid. *supra* note UN Security Council S/2010/91, Report of the Monitoring Group on Somalia submitted in accordance with resolution 1853 (2008), Paragraph 67 and 148, 2010.

[189] Ibid. *supra* note UN Security Council S/2010/91, Report of the Monitoring Group on Somalia submitted in accordance with resolution 1853 (2008), Paragraph 155–157, 2010.

In 2011, the Monitoring Group stressed the challenges with respect to information gathering from Eritrea due to restrictions on the movement of foreigners, including UN personnel and diplomats as well as the total absence of independent media or NGOs.[190] Due to such restrictions, the core of information gathering was diaspora sources and former Eritrean military, intelligence and diplomatic officials as well as monitoring of open sources in regional languages Somali, Tigrina and Arabic. Referring to arms trafficking and according to the available sources, Somalia had received at least 450 Man Portable Air Defence Systems (MANPADS) since the mid-1970s and part of them were in the hands of Al-Shabaab.[191]

In 2012, the Monitoring Group reported that the pattern of arms embargo violations was continuing, with Yemen being the principal source of supply of arms and ammunition for Somali non-state armed groups. A significant development was the large-scale and sustained development of foreign military forces in Somalia. Referring to the methodology, the challenges of gathering evidence persisted. The Coordinator of the Monitoring Group was declared *persona non-grata* and, in reply, retaliatory measures were considered to censure individuals who attempted to obstruct the work of the Group. Several confidential annexes were delivered to the Security Council Committee on Somalia.[192] A notable fact about new means of information was in Annex 6.2. when for the first-time the use of satellite imagery to identify specific evidence was mentioned.[193]

In 2013, the Monitoring Group acknowledged the extensive use of satellite imagery, giving examples of UN Organizations using evidence from satellites and their usefulness for gathering information in Somalia. In particular, individual UN agencies and NGOs had managed to set a higher standard of accountability through adoption of mechanisms to mitigate risks including direct monitoring, third party monitoring and remote sensing through satellite imagery.[194] Also, it was pointed out that the UN Food and Agriculture Organization (FAO) had started to systematically use satellite imagery for monitoring its infrastructure and rehabilitation in high-risk and inaccessible areas. Such evidence had helped FAO "to identify irregularities and establish further mechanism to address them".[195]

In 2014, evidence from satellite imagery was more noticeable in the Report of the Monitoring Group. Satellite imagery was not mentioned in the methodological

---

[190] UN Security Council S/2011/433, Report of the Monitoring Group on Somalia and Eritrea submitted in accordance with resolution 2002 (2011), Paragraph 10–11, 2011.

[191] UN Security Council S/2011/433, Report of the Monitoring Group on Somalia and Eritrea submitted in accordance with resolution 2002 (2011), Paragraphs 158, 2011.

[192] UN Security Council S/2012/544, Report of the Monitoring Group on Somalia and Eritrea submitted in accordance with resolution 2002 (2011), Paragraphs 1–7, 2012.

[193] Ibid. *supra* note UN Security Council S/2012/544, Report of the Monitoring Group on Somalia and Eritrea submitted in accordance with resolution 2002 (2011), pp. 311, 2012.

[194] UN Security Council S/2013/413, Report of the Monitoring Group on Somalia and Eritrea submitted in accordance with resolution 2060 (2012), Paragraph 13–14, 2013, https://www.undocs.org/S/2013/413, (accessed 26.05.2019).

[195] Ibid. *supra* note UN Security Council S/2013/413, Report of the Monitoring Group on Somalia and Eritrea submitted in accordance with resolution 2060 (2012), Paragraphs 13–14, 2013.

enunciation and was not used to detect the illicit trafficking of weapons, however, it supported the evidence of many activities. Satellite imagery was beneficial to verify project implementation and monitor potential diversion, one of the outcomes being the reduction of the list of partners for the Common Humanitarian Fund. Satellite imagery and GNSS supported the UN System in Somalia to improve coordination and increase internal transparency.[196] Using satellites for evidence provided the possibility of analysis of high-resolution satellite images of production areas, in particular charcoal production sites.[197]

In 2015, the Monitoring Group used satellite imagery to evidence production and export of charcoal in southern Somalia.[198] Also, satellite imagery was used to prove a missing office building for which the funds had been withdrawn from the Central Bank of Somalia (CBS).[199] The first time evidencing an attack, the Report presented the Garissa University College attack, in particular to evidence the route taken by the attackers.[200] Satellite imagery continued to evidence the movement of charcoal stockpiles.[201]

In the Report of 2016, the Monitoring Group continued to give examples of how satellite imagery supported evidence for the UN or NGOs for example, to monitor aid distribution,[202] charcoal stockpiles,[203] public land appropriation,[204] and land near Adde International Airport.[205] Satellite imagery was also used to analyse the

---

196 UN Security Council S/2014/726, Report of the Monitoring Group on Somalia and Eritrea submitted in accordance with resolution 2111 (2013), Paragraphs 38–39, 2014, https://www.undocs.org/S/2014/726, (accessed 26.05.2019).

197 UN Security Council S/2014/726, Report of the Monitoring Group on Somalia and Eritrea submitted in accordance with resolution 2111 (2013), Paragraphs 5-9, 18, 22, Annex 9.1 and Paragraph 24-31 Annex 9.1.a and Paragraph 32-27 Annex 9.1.b, 2014.

198 UN Security Council S/2015/801, Report of the Monitoring Group on Somalia and Eritrea submitted in accordance with resolution 2182 (2014), Paragraphs 150–151, 2015, https://www.undocs.org/S/2015/801, (accessed 26.05.2019).

199 Ibid. *supra* note UN Security Council S/2015/801, Report of the Monitoring Group on Somalia and Eritrea submitted in accordance with resolution 2182 (2014), Paragraph 30, 2015.

200 Ibid. *supra* note UN Security Council S/2015/801, Report of the Monitoring Group on Somalia and Eritrea submitted in accordance with resolution 2182 (2014), Paragraph 13 Annex 4.2, pp. 189, 2015.

201 Ibid. *supra* note UN Security Council S/2015/801, Report of the Monitoring Group on Somalia and Eritrea submitted in accordance with resolution 2182 (2014), Paragraph 1 Annex 8.1, pp. 300, 2015.

202 UN Security Council S/2016/919, Report of the Monitoring Group on Somalia and Eritrea submitted in accordance with resolution 2244 (2015), Paragraph 94, 2016, https://www.undocs.org/S/2016/919, (accessed 26.05.2019).

203 Ibid. *supra* note UN Security Council S/2016/919, Report of the Monitoring Group on Somalia and Eritrea submitted in accordance with resolution 2244 (2015), Paragraphs 130–131, 2016.

204 Ibid. *supra* note UN Security Council S/2016/919, Report of the Monitoring Group on Somalia and Eritrea submitted in accordance with resolution 2244 (2015), Paragraphs 43, Annex 4.6, 2016.

205 *Ibid.* supra note UN Security Council S/2016/919, Report of the Monitoring Group on Somalia and Eritrea submitted in accordance with resolution 2244 (2015), Paragraph 49, pp. 125, 2016.

Puntland Maritime Police Force and Puntland Security Forces base, highlighting the construction of new structures.[206]

In 2017, the Monitoring Group on Somalia and Eritrea issued the Report on Somalia for 2016, where the use of satellite imagery as evidence of the violation of the arms embargo on Somalia was underlined. In particular, satellite imagery acquired between December 2016 and September 2017 proved the establishment of a foreign military base in Berbera involving the transfer of military materiel to the territory which, in fact, constituted a violation of the embargo on Somalia.[207] Satellite imagery of the coast north of Berbera Airport between 23 December 2016 and 21 September 2017 showed easily detectable new facilities. In addition, the production, transport and stockpiles of charcoal in Somalia were evidenced. Illicit export continued at a high rate and according to the data analysed by the Monitoring Group based on satellite imagery in FAO, there were 26,000 total sites of charcoal production from 2011 to 2017 with 4,000 sites identified during 2017.[208] Satellite imagery of the stockpiles was useful for the analysis of the main sources of exports of charcoal. Satellite imagery showed charcoal stockpiles at Buur Gaabo, 9 July 2017 and also in northern and southern Kismayo on 24 June 2017.

Also, in 2017, the Report contained data about an attack that had occurred in Mogadishu consisting of two large vehicle-borne, Improvised Explosive Devices (IED), which exploded near the perimeter of the Adan Adde International Airport complex. The satellite imagery of the blast site was helpful in analysing the data and reporting about the incident.[209] In particular, it contributed to the analysis of the blast from an explosive device with estimated 1,200 kg TNT equivalence, catalogued as the largest IED by weight ever employed by Al-Shabaab.[210] The estimate was based on interviews with IED experts in Mogadishu as well as with an independent explosives engineer who analysed the blast scene using satellite imagery and crater dimensions.

In 2018, the Monitoring Group on Somalia and Eritrea included the analysis of satellite imagery in the methodology for evidentiary standards and the verification process.[211] In a continuation from the 2017 Report, it acknowledged the construction of a United Arab Emirates (UAE) military base in Berbera and correspondence

[206] Ibid. *supra* note UN Security Council S/2016/919, Report of the Monitoring Group on Somalia and Eritrea submitted in accordance with resolution 2244 (2015), Annex 8.9, 2016.

[207] UN Security Council S/2017/924*, Somalia Report of the Monitoring Group on Somalia and Eritrea submitted in accordance with resolution 2317 (2016), Para. 148–149, pp. 36/179, 2017. https://www.un.org/securitycouncil/sanctions/751/work-and-mandate/reports, (accessed 25.05.2019).

[208] Ibid. *supra* note UN Security Council S/2017/924*, Somalia Report of the Monitoring Group on Somalia and Eritrea submitted in accordance with resolution 2317 (2016), Paragraph 203–2015, pp. 46/179, 2017.

[209] Ibid. *supra* note UN Security Council S/2017/924*, Somalia Report of the Monitoring Group on Somalia and Eritrea submitted in accordance with resolution 2317 (2016), Annex 1.3., pp. 57/179, 2017.

[210] Ibid. *supra* note UN Security Council S/2017/924*, Somalia Report of the Monitoring Group on Somalia and Eritrea submitted in accordance with resolution 2317 (2016), pp. 10/179, 2017.

[211] UN Security Council S/2018/1002, Somalia Report of the Monitoring Group on Somalia and Eritrea submitted in accordance with resolution 2385 (2017), Paragraph 6, 2018.

with the Somaliland Administration regarding the violation of the arms embargo on Somalia. The establishment of the military base and expansion of port facilities, the latter being presented by the UAE within the scope of "… achieving security, stability and prosperity for the Republic of Somalia and its people" was evidenced by satellite imagery captured on 21 September 2018.[212] The Monitoring Group strongly praised the usefulness of satellite imagery by stating that "The most authoritative estimates regarding charcoal production in Somalia have been compiled by the Somalia Water and Land Information Management project of the Food and Agriculture Organization of the UN through satellite imagery analysis".[213]

Following from the 2017 Report findings about the largest IED in Al-Shabaab's history, the Monitoring Group detailed the formulae and tools used to estimate the explosion. Among the main evidence for the post-blast investigation, explosive engineering software was used to conduct the analysis, which included analysis of the damage radius from satellite imagery.[214] In particular, it contained evidence of a semi-circular pattern of damage, with buildings destroyed up to 100 m.[215] In the investigation, and in order to ascertain the level of damage, the Monitoring Group used diverse data, including the UN Institute for Training and Research (UNITAR) imagery analysis of the explosion, using the UN Operational Satellite Programme (UNOSAT), with the buildings marked as destroyed (red dot), severely damaged (orange dot) and moderately damaged (yellow dot).[216] Satellite imagery was also used to GPS track the route of the vehicle containing the bomb.[217]

The Monitoring Group used satellite imagery also to evidence several instances of forced evictions of internally displaced persons in Mogadishu, Baidoa, Bosaso, Galkayo and in the areas affected by the conflict between Somaliland and Puntland in the Sool region. It showed the scale of the eviction of an estimated 3,000 households as a result of Federal Government security forces using bulldozers to clear several camps for internally displaced persons.[218]

---

[212] Ibid. *supra* note UN Security Council S/2018/1002, Somalia Report of the Monitoring Group on Somalia and Eritrea submitted in accordance with resolution 2385 (2017), Paragraph 41–43, 2018.

[213] Ibid. *supra* note UN Security Council S/2018/1002, Somalia Report of the Monitoring Group on Somalia and Eritrea submitted in accordance with resolution 2385 (2017), Paragraph 170, 2018.

[214] Ibid. *supra* note UN Security Council S/2018/1002, Somalia Report of the Monitoring Group on Somalia and Eritrea submitted in accordance with resolution 2385 (2017), pp. 75/156, 2018.

[215] Ibid. *supra* note UN Security Council S/2018/1002, Source: UNMAS Somalia (2017) "Report on the VBIED Attacks in Mogadishu on 14 October 2017" pp. 76.

[216] Ibid. *supra* note UN Security Council S/2018/1002, Somalia Report of the Monitoring Group on Somalia and Eritrea submitted in accordance with resolution 2385 (2017), pp. 77, 2018.

[217] Ibid. *supra* note UN Security Council S/2018/1002, Somalia Report of the Monitoring Group on Somalia and Eritrea submitted in accordance with resolution 2385 (2017), pp. 87, 2018.

[218] Ibid. *supra* note UN Security Council S/2018/1002, Somalia Report of the Monitoring Group on Somalia and Eritrea submitted in accordance with resolution 2385 (2017), Paragraphs 155–157, 2018.

2.4.8 The Commission on Human Rights in South Sudan

The Commission on Human Rights in South Sudan submitted its first report in 2017. The Commission was established in March 2016, extended through 2017, and for a further year in March 2018 to monitor and report on the human rights situation in South Sudan. In reporting the facts, the Commission was mandated by the Human Rights Council to "collect and preserve evidence of, and clarify responsibility for alleged gross violations and abuses of human rights and related crimes".[219]

The 2019 Report of the Commission on Human Rights in South Sudan contains a summary of the events that occurred in 2018, on which the Commission collected and preserved evidence about violations that amount to international crimes, including war crimes and crimes against humanity.[220] The Commission conducted its work within the framework of international law, including international humanitarian law and international human rights law, and provided a summary analysis of all the emblematic incidents since 2013, focusing on the violations and crimes committed in 2018.[221] According to this Report, the Commission employed the best practices of fact-finding, being concerned also with assuring safety, security, confidentiality and well-being of all witnesses.[222] As the Commission underlined, such evidence is necessary "*with a view to ending impunity and providing accountability, and to make such information available also to all transitional justice mechanisms, including those to be established pursuant to Chapter V of the Agreement on the Resolution of the Conflict in the Republic of South Sudan, including the hybrid court for South Sudan, once established in cooperation with the African Union*".[223] The standard of evidence adopted by the Commission to support future accountability mechanisms, including criminal accountability, was based on "reasonable grounds to believe".[224]

This evidence, including the proof, resulted from the analysis of satellite imagery and was gathered by the Commission with the purpose to *make such information available to all transitional justice mechanisms*. The standard of proof has to be consistent with the practice of other UN fact-finding bodies, thus employing a "reasonable grounds to believe" necessary for factual determinations on individual cases, incidents and patterns of conduct. In this regard, an example of how satellite images

---

219 A/HRC/RES/37/31 Human Rights Council, Situation of human rights in South Sudan, 13 April 2018, https://www.securitycouncilreport.org/atf/cf/%7B65BFCF9B-6D27-4E9C-8CD3-CF6E4FF96FF9%7D/a_hrc_res_37_31.pdf, (accessed 24.09.2019).

220 A/HRC/40/69, Report of the Commission on Human Rights in South Sudan, 2019, https://www.ohchr.org/EN/HRBodies/HRC/CoHSouthSudan/Pages/Index.aspx, (accessed 01.06.2019).

221 A/HRC/40/69, Report of the Commission on Human Rights in South Sudan, Paragraph 62, 2019, https://www.ohchr.org/EN/HRBodies/HRC/CoHSouthSudan/Pages/Index.aspx, (accessed 01.06.2019).

222 Ibid. *supra* note A/HRC/40/69, Report of the Commission on Human Rights in South Sudan, Paragraph 13, 2019.

223 Ibid. *supra* note A/HRC/40/69, Report of the Commission on Human Rights in South Sudan, Paragraph 8, 2019.

224 Ibid. *supra* note A/HRC/40/69, Report of the Commission on Human Rights in South Sudan, Paragraph 11, 2019.

were used for analysis to provide evidence to the Commission is the revealing of the damage suffered by approximately 7,345 structures. The damage resulted in massive population displacement and food insecurity, but also a lack of assistance, leading in some cases to the withdrawal of humanitarian organizations from the fighting zones.[225]

Despite the "Cessation of Hostilities Agreement of December 2017, the Khartoum Declaration of June 2018, the permanent ceasefire and the Revitalized Agreement of September 2018" on which the commission had collected and preserved evidence, the Commission further registered attacks on civilians, in violation of human rights.[226] The atrocities against the civil population, including continuous and violent attacks and their consequences were proved also via satellite imagery. The analysis provided the Commission with evidence and established how many structures were damaged or destroyed and the violations of the soldiers who intended to destroy local crops.[227] The renewed fighting and continuing violation of human rights in South Sudan was documented by the Commission, which has gathered evidence for future accountability processes.[228] The confidential dossier will be handled by the Commission to the UN High Commissioner for Human Rights in Geneva. "The evidence we have collected and preserved will be available to the prosecutor of the future Hybrid Court and other transitional justice mechanisms" and "This evidence may be used beyond South Sudanese bodies—it may be available on request to regional and state parties for future prosecutions".[229]

In conjunction with the report of the Commission on Human Rights in South Sudan (A/HRC/40/69) the conference paper reflects in greater detail the evidence that was collected and its findings. As sources of direct information, the Commission included "Satellite imagery analysis from reliable sources, authenticated video and photo material" in addition to evidence from interviews of victims and eyewitnesses, medical examinations, publicly available admissions, policies and directives of the Government of South Sudan, statistics, surveys and situation reports.[230] Satellite imagery provided by UNOSAT was in particular used to analyse the Mayemdit and Leer areas as stated by the Commission in the "UNOSAT analysis of destruction in

---

[225]Ibid. *supra* note A/HRC/40/69, Report of the Commission on Human Rights in South Sudan, Paragraph 71, 2019.

[226]Ibid. *supra* note A/HRC/40/69 Report of the Commission on Human Rights in South Sudan, Paragraph 36, 2019.

[227]Ibid. *supra* note A/HRC/40/69, Report of the Commission on Human Rights in South Sudan, Paragraph 88, 2019.

[228]UN Commission on Human Rights in South Sudan News, Outraged by renewed fighting and continuing human rights violations in South Sudan, UN Human Rights Experts urge all parties to stop conflict, end impunity and respect provisions of the revitalized peace agreement, 20 February 2019, https://www.ohchr.org/EN/HRBodies/HRC/Pages/NewsDetail.aspx?NewsID=24183&LangID=E, (accessed 04.06.2019).

[229]Ibid. *supra* note UN Commission on Human Rights in South Sudan News, 20 February 2019.

[230]A/HRC/40/CRP.1, CRP on the Report of the Commission on Human Rights in South Sudan, Paragraph 17, 2019, https://undocs.org/pdf?symbol=en/A/HRC/40/CRP.1, (accessed 01.06.2019).

Western Bahr el Ghazal State and Unity State", South Sudan, 23 January 2019, ERN D116708-D116714.[231]

The analysis of evidence established that fighting and attacks resulted in massive displacement.[232] As the Commission acknowledged, it received the assistance and contributions of UN agencies, civil society organizations and experts.[233] The satellite imagery analysis was provided by UNOSAT and described the damage to or destruction of the structures.[234] The analysis was useful for the Commission to focus not only on establishing the occurrence of violations but also identifying those responsible for the violations and crimes.[235] The evidence collected provided "reasonable grounds" to prove violations on human rights using the standard "to believe that an incident or pattern of conduct has occurred only when it has obtained a reliable body of information, consistent with other material, upon which a reasonable and ordinarily prudent person would believe that the incident or pattern of conduct has occurred".[236] Evidence was prepared to be provided or to be used by various accountability mechanisms such as the Hybrid Court for South Sudan, these conditions implementing a standard to support a future accountability mechanism.[237]

Finally, the UN Human Rights Council organized a working session in Addis Ababa on strategies for evidence collection. The working agenda included discussion on possible improvements in evidence collection for criminal accountability. In the background information the gap between reports of human rights violations, in particular sexual violence, and the number of formal prosecutions, was underlined. Taking Sexual and Gender-Based Violence (SGBV) in South Sudan as an example, the Commission noted that proper documentation and investigation is critical for the success of any future criminal prosecutions and transitional justice mechanisms. It recalled that international tribunals have on many occasions pointed out the insufficiency or even the inadmissibility of evidence coming from third parties, or dismissed

---

231 Ibid. *supra* note A/HRC/40/CRP.1, CRP on the Report of the Commission on Human Rights in South Sudan, Paragraph 715, 2019 https://undocs.org/pdf?symbol=en/A/HRC/40/CRP.1, (accessed 01.06.2019).

232 Ibid. *supra* note A/HRC/40/CRP.1, CRP on the Report of the Commission on Human Rights in South Sudan, Paragraph 717, 2019, "*In April alone, 7,000 people were displaced in and around Koch town. In May, 40,000 people were reported to have been displaced in Leer County, out of a total population of approximately 80,000. This, at a time when the Integrated Food Security Phase Classification system predicted over 70 percent of Unity State to be at Crisis, Emergency or Catastrophe (Phase 3-5) levels of food insecurity*".

233 Ibid. *supra* note A/HRC/40/CRP.1, CRP on the Report of the Commission on Human Rights in South Sudan, Paragraph 36, 2019.

234 Ibid. *supra* note A/HRC/40/CRP.1, CRP on the Report of the Commission on Human Rights in South Sudan, Paragraph 815, 2019.

235 Ibid. *supra* note A/HRC/40/CRP.1, CRP on the Report of the Commission on Human Rights in South Sudan, Paragraph 32, 2019.

236 Ibid. *supra* note A/HRC/40/CRP.1, CRP on the Report of the Commission on Human Rights in South Sudan, Paragraph 14 and examples at Paragraphs 854, 855, 864, 2019.

237 Ibid. *supra* note A/HRC/40/CRP.1, CRP on the Report of the Commission on Human Rights in South Sudan, Paragraph 13, 2019.

the prosecution's arguments to establish criminal responsibility. It is to be mentioned that in the overall process of gathering evidence, satellite-based imagery can support enhancing documentation and evidence collection, thus making a contribution to more effective prosecutorial systems for future justice mechanisms.[238]

Overall, the mechanism to collect, consolidate, preserve and analyse evidence by the Commission should facilitate fair and independent criminal proceedings. Because satellite imagery analysis is capable of complying with international law standards, in national, regional or international courts, the Commission on South Sudan has collected evidence of human rights violations and abuses. Satellite images are capable of wide area monitoring, and in particular are very helpful for surveillance in remote areas; satellites can operate day and night, and the collection of images can be then compared during a certain period of time, which is much more effective for investigations than only after the incident footage; the content can be effectively shared, accessed and used to ensure a large and continuously increasing amount of evidence.

2.5 Natural Disasters

2.5.1 Weather and Atmospherically Monitoring

Monitoring weather from space can provide timely and accurate data that can then be used for monitoring earth weather and analysing climate studies. Different information such as atmospheric temperature profiles, relative humidity profiles, fractional cloud cover, cloud pressure and temperature; the total ozone burden; mapping of the distribution of atmospheric gases; sea surface temperature; surface albedo; snow and ice cover; outgoing long wave radiation; and the precipitation index can all be measured from space.[239] EO satellites have proven to be vital tools for weather and atmospheric monitoring. Analysing such important information is necessary for numerous questions to be addressed in Earth Science such as:

(A) Cloud-radiation interaction
(B) Global energy and water cycles
(C) Measures of the increase of the greenhouse gases
(D) Atmosphere-surface interaction
(E) Land properties
(F) Improvement of weather forecasting

[238]Prosecuting Sexual and Gender-Based Crimes in South Sudan: working session on strategies for evidence collection, Addis Ababa 4–5 March, https://www.ohchr.org/EN/HRBodies/HRC/CoHSouthSudan/Pages/Index.aspx, (accessed 04.06.2019).

[239]Chedin, A., Chahine, M.T., Scott, N.A., High Spectral Resolution Infrared Remote Sensing for Earth's Weather and Climate Studies, pp. VI, Springer, 1993.

2.5.1.1 Monitoring Weather

When it comes to weather, people on Earth are directly affected by the force of nature. In recent years weather has become increasingly unstable, thus the challenge is not only to mitigate harmful action and adapt to climate change but also to be able to predict and forecast. All sectors rely on accurate forecasts for energy, transportation, construction, agriculture, tourism and utilities. Weather forecasts are necessary both for safety of the people involved in various travelling activities and to prepare against floods, heavy precipitation or heat waves but also for the economy, especially agriculture, and to reduce damage to property and infrastructure.

Observations from satellites have become indispensable for forecasting the weather. For example, to issue warnings on rapid developing phenomena, "nowcasting" is real-time detection. This process requires high-quality images of the atmosphere delivered in real time from space. According to EUTESTAT, its geostationary satellites such as METEOSAT have the capability to deliver real time imagery every 5 to 15 min, representing a valuable input of data. Also, satellites provide a global view from the polar orbit, at approximately 800 km from the Earth and the METEOP Satellites in Polar Region of EUMETSAT are used to forecast weather up to 10 days in advance.

2.5.1.2 Monitoring Climate

A better understanding of the impact of climate change could be provided to governments and decision makers from space. Meteorological satellite programs are a valuable asset for climate monitoring and their role will grow. There is more than 30 years of experience in working with meteorological data and further international cooperation is needed such as reprocessing data, extracting climate records, and making such data available to downstream applications and scientists. Meteorological satellites can provide useful data for climate monitoring. Satellite observations are fundamental to analyse the layer of ozone and source methane, carbon monoxide, aerosols, sulphur compounds and nitrogen oxides. Within the Global Framework for Climate Services (GFSC) established by the World Meteorological Organization, climate services are based on a combination of observations, climate prediction and projections at longer scales. The so-called requirements for observations of Essential Climate Variables (ECV) are established and maintained by the Global Climate Observing System (GCOS) program.[240] The EUMETSAT role in climate monitoring activities for example, provides calibration products to improve the consistency of the data received from various satellites, and is a core contributor to the Global Space-based Inter-Calibration Systems (GSICS). Accordingly, the methodologies

[240]EUMETSAT, Climate services need to be developed to meet the challenge of mitigating and adapting to climate change, https://www.eumetsat.int/website/home/AboutUs/WhatWeDo/MonitoringClimate/index.html, (accessed 10.08.2019).

developed in GSICS are applied to the reprocessing of historical data.[241] In relation to the changing climate, satellites can observe “systematically and globally” many of the essential climate variables (ECV) identified by the GCOS. Satellites are an essential architecture useful for climate monitoring from outer space.

2.5.1.3 Monitoring Oceans

The oceans cover about 71 percent of the Earth’s surface and contain 97 percent of the Earth’s water. Thus, it is impossible to make a global analysis only from in situ observations. Satellites can provide a trusted source of highly accurate global measurements of the physical and biological state of the ocean including sea level, ocean current, sea surface temperature, ocean colour and sea ice. Monitoring oceans involves gathering data from space about the seas and ocean, at global, basin and coastal scales.[242] Application areas for remote sensing include safety, shipping, fisheries, management of marine resources, coastal ecosystems and marine pollution, offshore industry as well as ice and surface waters.

2.5.1.4 Atmospheric Composition

Air pollution is responsible for a broad range of adverse health effects. Environmental protection policies are implemented globally, with each state finding its own solutions. Air quality forecasts are based on Numerical Weather Prediction (NWP) models. Analysis is based on observations of trace gases and particles, taking into account natural and anthropogenic emissions. The operational monitoring and forecasting of atmospheric composition can be made through remote sensing from space. EUMETSAT for example, uses its geostationary and polar-orbiting satellites for monitoring atmospheric composition, including volcanic ash and dust storms or ozone, NO2, CO, CH4 in the ultraviolet and thermal infrared parts of the spectrum.[243] Monitoring ozone depth can be performed through satellite data.[244] The ozone layer has been measured through satellites since 1978, which was essential in documenting and

[241] Ibid. *supra* note EUMETSAT, Climate services need to be developed to meet the challenge of mitigating and adapting to climate change. https://www.eumetsat.int/website/home/AboutUs/WhatWeDo/MonitoringClimate/index.html, (accessed 24.09.2019).

[242] EUMETSAT supports the development of operational oceanography through the delivery of ocean data to its Member States, the Copernicus marine service, and users worldwide, https://www.eumetsat.int/website/home/AboutUs/WhatWeDo/MonitoringOceans/index.html, (accessed 24.09.2019).

[243] Observations of EUMETSAT satellites are critical inputs to monitoring and forecasts of air quality, which are increasingly important for the health of European citizens, https://www.eumetsat.int/website/home/AboutUs/WhatWeDo/AtmosphericComposition/index.html, (accessed 24.09.2019).

[244] Calle, A., Casanova, J.L., Ozone in the Atmosphere, in Chuvieco, E. (Eds.), Earth Observation of Global Change, The Role of Satellite Remote Sensing in Monitoring the Global Environment, pp. 59–84, Springer Netherlands, 2008.

discovering the ozone layer hole in 1985. Different satellites were used to monitor the dynamic of the ozone on the stratosphere and troposphere.[245]

2.5.1.5 Copernicus

COPERNICUS is the European system for monitoring the Earth and is coordinated and managed by the European Commission. In particular, the observation infrastructure is performed under ESA for the space component and by the European Environment Agency and EU countries for the in situ component.[246] As a joint initiative of the EU and ESA, it aims to provide monitoring of the atmosphere, oceans and continental surfaces. COPERNICUS covers services related to atmosphere, marine, land, climate change, security and emergency. The COPERNICUS Space infrastructure and the COPERNICUS services are in their vast majority accessible to any citizen and organization on a free, full and open access basis.[247] In order to facilitate and standardize access to data, COPERNICUS users can use five cloud-based platforms known as Data and Information Access Services (DIAS) with access to COPERNICUS data and information. The cloud-based tools are open source and/or on a pay-per-use basis.[248] Access points to COPERNICUS satellites are managed by ESA (two): respectively COPERNICUS Open Access Hub and COPERNICUS Space Component Data Access (CSCDA); and also by EUMETSAT (two): EUMETCast and COPERNICUS Online Data Access (CODA).[249]

---

[245] Ibid. *supra* note Calle, A., Casanova, J.L., Ozone in the Atmosphere, in Chuvieco, E. (Eds.), Earth Observation of Global Change, The Role of Satellite Remote Sensing in Monitoring the Global Environment, pp. 59–84, Springer Netherlands, 2008.

> since 1978 sensor TOMS, originally on-board Nimbus-7 and later on satellites like the Russian Meteor-3 or the Japanese Adeos, has been the one mostly used for analyses on a large scale completing measurement series. The NOAA series satellites also include a sensor called HIRS with an infrared band centred in an ozone absorption peak at 9.6 μm. This band has been included in the current MODIS sensor and in the European geostationary MSG. GOME sensor of the European satellite ERS has been very important in the continuation of data series and, at this moment, we should mention the relevance of the European satellite ENVISAT, which carries GOMOS (Global Ozone Monitoring by Occultation of Stars) among other sensors. GOMOS is able to generate stratospheric ozone profiles with a high accuracy above 15 km from the measurement of a stellar spectrum when it is visualised from the space through the earth's atmosphere.

[246] Copernicus, https://ec.europa.eu/growth/sectors/space/copernicus_en, (accessed 24.09.2019).

[247] Copernicus, Access to Data, https://www.copernicus.eu/en/access-data, (accessed 24.09.2019).

[248] Copernicus, DIAS, https://www.copernicus.eu/en/access-data/dias, (accessed 24.09.2019).

[249] Copernicus, Conventional Data Access Hubs, https://www.copernicus.eu/en/access-data/conventional-data-access-hubs, (accessed 24.09.2019).

2.5.1.6 The National Oceanic and Atmospheric Administration

The National Oceanic and Atmospheric Administration (NOAA) manages data from a constellation of geostationary and polar-orbiting meteorological satellites. Remote sensing is used to provide various geophysical variables that can be useful in analysing the Earth's atmospheric, oceanic, and terrestrial domains.[250] As an example of NOAA EO activity, satellites help monitor and predict tropical climate events such as tornadoes, floods and dust storms but also volcanic eruptions and forest fires, mostly done by analysing data from geostationary satellites. Also, polar orbiting satellites can monitor precipitations, sea surface temperatures, atmospheric temperature and humidity, ice extinction on sea and oceans and global vegetation analysis including forest fires and volcanic eruptions.

2.5.1.7 Eumetsat

The European Organization for the Exploitation of Meteorological Satellites (EUMETSAT) is an intergovernmental organization. According to its convention, the primary objective of EUMETSAT is to establish, maintain and exploit the European system of operational meteorological satellites, and also to contribute to operational monitoring of the climate and the detection of global climatic changes.[251] Observations from EUMETSAT satellites increase the performance of weather forecasts and are also used to produce climate records of solar radiation parameters. As part of international cooperation, since 1987 EUMETSAT has served as the Secretariat for the Coordination Group for Meteorological Satellites (CGMS), which is a platform for exchange of technical information on geostationary and polar orbiting satellite systems. The CGMS Secretariat represents CGMS members at the Committee on EO Satellites (CEOS), the Group on EO (GEO) and the Space Frequency Coordination Group (SFCG).[252]

2.5.2 *Toxic Waste Identification and Documentation*

The global waste market is valued in $ billions. Illegal activities have flourished by exploiting the high costs of legal waste management and disposal and evading tax and other costs associated with disposing waste. The environmental consequences

---

[250] NOA website, Satellite Data, https://www.ncdc.noaa.gov/data-access/satellite-data, (accessed 09.08.2019).

[251] Convention for the Establishment of a European Organization for the Exploitation of Meteorological Satellites (EUMETSAT), entered into force on 19 June 1986, including amendments entered into force on 19 November 2000.

[252] EUMETSAT, International Cooperation, https://www.eumetsat.int/website/home/AboutUs/InternationalCooperation/CoordinationGroupforMeteorologicalSatellitesCGMS/index.html (accessed 24.09.2019).

include loss of habitat, and pollution of groundwater, but economic consequences such as landfill tax and loss of recycling profit are equally significant. By offering cheaper unlawful landfills, the danger is to harm environment, wildlife and natural resources. Non-compliant operators are to be brought to courts that have to assess the volume of waste, source of waste such as toxic, domestic or demolition and whether it has polluted areas outside of the owner's land. Both geophysics and space remote sensing from space are a cost-effective solution that could assist courts in answering these questions.[253]

Toxic waste includes nuclear radioactive waste that in case of inappropriate management or illegal disposal could result in serious biosphere implications. The need to correctly identify, evaluate and choose areas that can host a permanent disposal facility has serious political and environmental implications. Satellite technology could contribute to better decision making in repository site selection. For example, public discussions have revealed that beneath Greenland's icy surface there is toxic nuclear waste, left over by a U.S. military research base "Camp Century" from the Cold War and which may eventually be exposed due to climate change that is melting the ice at a rapid pace.[254] This is the most recent example that shows the need to continuously manage toxic waste and find solutions for the long term.

To select a geological disposal site for high-level radioactive waste, governments have enacted laws and regulations to specify the procedure for repository site selection. For example, Japan has released a recent study to assess the suitability and applicability of the Japanese site screening process for identifying the areas for geological disposal of toxic waste in comparison with international best practices. During the working group assessment, the 2011 Fukushima Daiichi nuclear power plant accident and the need for technical safety basis of geological disposal facilities, including the long-term safety of geological disposal, were recalled.[255]

Satellite imagery could be a useful tool in waste management. The ESA Business Application has underlined that space added value implies using EO data for

---

[253] Ruffel, A., Waste Management and disposal: locating illegal buried waster using remote sensing (satellites, aircraft, drones and geophysics), 18 May 2016, https://kess.org.uk/2016/05/18/waste-management-disposal-locating-illegal-buried-waste-using-remote-sensing-satellites-aircraft-drones-geophysics/ (accessed 18.08.2019).

[254] Colgan, J.D., Trump wants to buy Greenland. He might want to know about the toxic nuclear waste buried in its ice, The Washington Post 16.08.2016, https://www.washingtonpost.com/politics/2019/06/11/congress-worries-that-climate-change-will-force-us-pay-foreign-governments-millions-heres-why/?noredirect=on, (accessed 18.08.2019).

[255] OECD—Radioactive Waste Management, Japan's Siting Process for the Geological Disposal of High-level Radioactive Waste, An International Peer Review, pp. 9, 2016.

monitoring and GNSS for tracking illegal dumping activities.[256] The use of satellite imagery can identify sites that may be appropriate for landfill and provide data about their development over time. Such data implies that the procedure to find a site would be respected and would be part of monitoring of the location, the areas and historical changes. On the other hand, satellite imagery could be used to detect any authorized changed in site structure through visual interpretation and AI procedures. Satellite imagery could also contribute to the study of the hydrological characteristics of soil. Finally, space sensors could provide important information about methane from landfill sites. Monitoring and assessment should make use of satellite data in addition to other techniques.[257]

Illegal waste sites are becoming a growing problem globally.[258] The increasing volume of waste generated from various activities raises questions about waste disposal problems and how to solve them. Satellite images could provide accurate identification of contamination and illegal landfill sites. Satellite remote sensing was identified as an ideal solution to identify illegal waste disposal sites if using high-resolution imagery. Satellite imagery could support efforts to control waste disposal, especially using satellite technology to monitor landfill sites and identify illegal waste disposal sites.[259] Thus, existing remote sensing methods and sensors could be used to monitor and map illegal waste disposal sites.

---

[256] ESA, WASTE IAP—Space based Support Services for Waste Management, Feasibility Study, 23 July 2018, https://business.esa.int/projects/waste-iap (accessed 18.08.2019).

Earth Observation data for monitoring:

Change detection routines are applied using multitemporal SAR data to identify potential sites where illegal waste tipping is occurring.

Optical data is used to both identify illegal waste sites and to validate the sites identified in the SAR data.

GNSS for Tracking:

Tracking devices using the location information from GNSS satellites inform when a waste consignment may be routed from its intended destination to an illegal dumping site

By combining this information with vehicle tracking, a full picture of unexpected behavior in the lifecycle of waste disposal is provided.

[257] AHRC—Arts & Humanities Research Council, Satellite Monitoring as a Legal Compliance Tool in the Environment Sector, Case Study Two: The Disposal of Waste on Land and Remote Sensing, Chapter 3 to Chapter 4, http://www.envirosecurity.org/helf/UCL%20Case%20Study%20on%20Waste%20Monitoring.pdf (accessed 18.08.2019).

[258] Purdy, R., The Eye in the sky that can spot illegal rubbish dumps from space, http://theconversation.com/the-eye-in-the-sky-that-can-spot-illegal-rubbish-dumps-from-space-98395 (accessed 24.09.2019).

[259] Ibid. *supra* note AHRC—Arts & Humanities Research Council, Satellite Monitoring as a Legal Compliance Tool in the Environment Sector, Case Study Two: The Disposal of Waste on Land and Remote Sensing, chapter 3 to 4,

http://www.envirosecurity.org/helf/UCL%20Case%20Study%20on%20Waste%20Monitoring.pdf, (accessed 24.09.2019).

2.5.3 *Marine Oil Spills*

As a result of fraudulent water management, the pollution of oceans is increasingly becoming a global problem, with direct consequences, such as damaging the marine environment, and indirect consequences for humans who depend on fish consumption. Fossil fuel oils and palm oil spills have a negative impact on the marine environment. They deplete water of oxygen, killing fish and coating the wings of birds, among others. Oil spills are highly dangerous and costly to reduce the impact on the environment once the damage has been done. The clean-up costs for oils spills are on average Euro 120 m each year. It has been estimated that 457,000 tons of oil escape the ocean every year, of which 50,000 tones is in Europe alone.[260] The vessels in violation are liable and can be fined up to several million US dollars, and crewmembers can be sentenced to prison. In addition, violators may face additional charges of obstruction of justice and the owners may be subject to strict compliance programs and banned from trading in certain waters.[261]

The effectiveness of satellites for oil spill surveillance has been proven many times. Radar imaging satellites are also used to observe the surface during all weather conditions, day and night, which is particularly useful in cloudy locations or from high altitudes.[262] For example, it was stated that ships are regularly dumping large quantities of palm oil off the coast of Britain, contaminating British waters. Such illegal actions practices have been documented by satellites operated by the European Maritime Safety Agency (EMSA) with oil slicks stretching in the southern part of the North Sea.[263]

Also, following two massive oil spills in 2008 and 2009, the SHELL company faced formal Court proceedings in the UK. In the preliminary issues hearing from 2014, the London High Court ruled that SHELL Nigeria could be legally liable for bunkering, it being argued that SHELL should have taken measures to protect its pipes infrastructure given the foreseeable risk of bunkering and illegal hacking into pipelines to steal the oil.[264] In January 2015, the Bodo Community in the Niger Delta achieved an historic victory when SHELL announced a £ 55 million settlement agreement in respect of two operational oil spills from pipelines in 2018. On the issue

---

260 Geocento, Oil Spill Monitoring, Satellites help to track the extent and spread of oil spills, https://geocento.com/satellite-imagery-case-studies/radar-imaging-satellites-for-oil-spills/, (accessed 19.08.2019).

261 Loss Prevention Briefing, Bilge Water Management, http://www.nepia.com/media/869523/Bilge-Water-Management-LP-Briefing.PDF, (accessed 19.08.2019).

262 Ibid. *supra* note Geocento, Oil Spill Monitoring, Satellites help to track the extent and spread of oil spills, https://geocento.com/satellite-imagery-case-studies/radar-imaging-satellites-for-oil-spills/, (accessed 24.09.2019).

263 Crisp, W., Ships dumping noxious palm oil off British coast without legal consequences, The Telegraph, 3 February 2018, https://www.telegraph.co.uk/news/2018/02/03/ships-dumping-noxious-palm-oil-british-coast-without-legal-consequences/, (accessed 19.07.2019).

264 LeighDay, London High Court rules that Shell Nigeria could be legally liable for bunkering, 20 June 2014, https://www.leighday.co.uk/News/2014/June-2014/London-High-Court-rules-that-Shell-Nigeria-could-b, (accessed 20.08.2019).

of the volume of oil spilt in the Niger Delta, SHELL used satellite imagery as evidence "the issues of the volume of oil released and the extent of the impact as a result of the 2008 oil spills will be the subject of expert evidence which will include hydrological data, satellite imagery and further data samples taken from the relevant areas".[265]

Also related to the Niger Delta, there have been three cases against SHELL in the Netherlands and UK. The first case was the Nigerian farmers versus SHELL. In 2008, four Nigerian farmers sued SHELL because of oil pollution in three Nigerian villages. They demanded SHELL to clean up the oil spills and pay compensation for the damages caused, but also to improve the maintenance of its pipelines and installations to prevent future spills. The investigation focused on establishing whether the oil spills were a result of sabotage or poor maintenance.[266] Another case was Okpabi versus SHELL, brought in front of a UK court in 2016, which involved oil spills contamination. For these cases and others, the satellite imagery is helpful to understand the possible damage and the cleaning efforts related to oil pollution.

ESA supplied satellite imagery, particularly SAR images, on a Spanish oil spill under the International Charter on Space and Major Disasters. The SAR images were requested by the European Commission on the basis of the International Charter on Space and Major Disasters. The Charter provides measures that allow satellite technology to be put at the service of rescue authorities. Under this agreement, ESA, CNES, the Canadian Space Agency (CSA), the Indian Space Research Organization (ISRO), and the US National Oceanic and Atmospheric Administration (NOAA) will contribute with satellite imagery upon request from authorized users. In particular, the satellite imagery was requested to identify the affected areas.[267]

Offshore oil and gas platforms are used for the extraction, processing and temporary storage of crude oil and natural gas in the sea.[268] The types of offshore oil and gas production structures are classified as: (i) floating offshore production systems; (ii)

---

[265] Amnesty International, Briefing for Investors: Shell's Growing Liabilities in the Niger Delta. Lessons from the Bodo Court, https://www.amnesty.org.uk/files/amnesty_international_briefing_on_shell_for_investors.pdf, (accessed 20.08.2019). See also: Paragraphs 7.4 and 34.4 of the document entitled: Celina Naadueba & Others and Shell, available at https://www.amnesty.org/en/latest/news/2014/11/court-documents-expose-shell-s-false-claims-nigeria-oil-spills/, Paragraphs 6.4 and 33.3 of the document entitled: Kembe Sangbara & Others Residing in Lewe and Shell.

[266] Greenpeace, Investor Briefing May 2018, Seeking justice: the rising tide of court cases against Shell, http://www.greenpeace.nl/Global/nederland/report/2018/klimaat_en_energie/Investor%20Briefing%20Seeking%20justice%20the%20rising%20tide%20of%20court%20cases%20against%20Shell.pdf, (accessed 21.08.2019).

[267] ESA, Earth Online Sample Data, https://earth.esa.int/web/guest/data-access/sample-data/-/asset_publisher/tg8V/content/prestige-oil-spill-galicia-spain-1623?p_r_p_564233524_assetIdentifie, (accessed 19.08.2019).

[268] Liu, Y., Sun, C., Sun, J., Li, H., Zhan, W., Yang, Y., Zhang, S., Satellite data lift the veil on offshore platforms in the South China Sea. Scientific Reports, 2016, https://doi.org/10.1038/srep33623, https://www.researchgate.net/publication/308365780_Satellite_data_lift_the_veil_on_offshore_platforms_in_the_South_China_Sea, (accessed 22.05.2019).

tension leg platforms; (iii) spar platforms; (iv) subsea production systems.[269] Common for many of these facilities is their location in remote, hostile, environmentally and potentially culturally sensitive locations. Energy companies regularly monitor sites for oil spills and leaks in addition to geological interpretation and assessment and seismic planning. The costs associated with conventional monitoring techniques could be significantly reduced if more of such observations were made using satellites and the interpretation of satellite-based imagery. In particular, satellites have relevant applications to offshore operations such as monitoring the environmental risk associated with routine operations “from the first seismic study commissioned for an oil or gas prospect to the final day of a field’s decommissioning”.[270] In addition, the offshore and near-shore infrastructure and facilities are vulnerable to malicious acts, which is why satellite imagery could be a useful tool in response to the many potential emergency scenarios, from terrorist attacks to accidental oil spillage.[271]

Satellite-based pollution monitoring can be performed regularly, as was demonstrated in a pilot test by AIRBUS Space and Defense over offshore Angola. The optical images from the Pleiades satellite were used to identify ship traffic and classify the types of ships entering the area around the production vessels.[272] Imagery used for oil and gas applications “range from around 0.5 m resolution for detailed studies to 15 m resolution for more regional applications”. Also, the most accurate technology can perform “a global elevation model at 12 m grid spacing, allowing more subtle terrain features to be identified”. In supporting oil and gas exploration, the value of satellite based remote sensing activities is in the “interpretation and analysis applied to the base imagery and elevation models, including geological interpretation and seismic planning”.[273] In addition to supporting exploration, monitoring includes security concerns where optical satellite imaging technology could be a viable solution. The latest satellite technology allows immediate tasking of a satellite to rapidly capture images of an emergency situation in any location around the globe. A fast reaction capability allows companies to initiate critical emergency action plans when dealing with urgent situations that endanger people, assets or the environment. Common security concerns vary from terrorist attacks to the illegal tapping of pipelines and, since oil and gas operations are increasingly operating in

[269] Singh, B., Different Types of Offshore Oil and Gas Production Structures, 5 August 2019, https://www.marineinsight.com/offshore/different-types-of-offshore-oil-and-gas-production-structures/, (accessed 22.05.2019).

[270] Hall, M., Airbus Defence and Space. Enhanced satellite imagery supports offshore oil spill, security monitoring, 5 February 2016, https://www.offshore-mag.com/production/article/16754876/enhanced-satellite-imagery-supports-offshore-oil-spill-security-monitoring, (accessed 22.05.2019).

[271] Ibid. *supra* note Hall, M., Airbus Defence and Space, 2016.

[272] Ibid. *supra* note Hall M., Airbus Defence and Space, 2016.

[273] Hall, M., Supporting Exploration with Satellite Imaging Technology, in Airbus News, 2017, http://www.airbusds-ogm.com/news/supporting-exploration-with-satellite-imaging-technology, (accessed 22.05.2019).

areas with security concerns, geospatial information provides a unique tool to monitor assets.[274] Thus, high resolution satellite imagery can be applied to oil and gas projects.[275]

2.5.4 International Charter Space and Major Disasters

Through global collaboration, Members of the Charter support national needs and make satellite data available in support of disaster management. Disaster situations require rapid response and different space agencies mobilize themselves by combining remote sensing data to provide as much accurate data as needed, without any cost for the user. In addition, various national and regional disaster monitoring organizations support international cooperation as co-operating bodies. From the moment the Charter was activated, space agencies and co-operating bodies have made available their expertise and the satellite data to be coordinated for rapid response to major disaster situations. In numbers, there have been 615 activations up to August 2019, 125 countries, 17 charter members and 61 contributing satellites.[276] Authorized users should have a national mandate to coordinate emergency response measures and be trained to download and access the maps.

Authorized Users must:

- *Be a national disaster management authority or its delegated agency in that country*
- *Have the capacity to download and utilize maps*
- *Be able to submit and pursue activation requests in English.*

There are different types of disasters that the Charter covers such as: (i) cyclones; (ii) earthquakes; (iii) fires; (iv) floods; (v) snow and ice; (vi) ocean waves; (vii) oil spills; (viii) volcanoes; (ix) landslides; (x) other. Regarding oil spills, the Charter underlines the challenges in locating oil spills as they are relatively small patches of oil in vast oceans, and the effect of ocean tide, currents and weather on the movement of oil spills, making them unpredictable.[277] Different product examples exist at international level where the charter has been activated for oil spills. The 2017 Report

---

[274] Ibid. *supra* note Hall, M., Supporting Exploration with Satellite Imaging Technology, in Airbus News, 2017, http://www.airbusds-ogm.com/news/supporting-exploration-with-satellite-imaging-technology/, (accessed 22.05.2019).

[275] Barnes, M., High Resolution Satellite Imagery Applied to Oil and Gas Projects, 2005 SEG Conference—Special Session, https://www.apsg.info/resources/Presentations/Misc/2005SEGConference_2005.11_MichaelBarnes_HighResolutionSatelliteImaging_OilandGas.pdf, (accessed 19.07.2019).

[276] International Charter Space & Major Disasters, About the Charter, https://disasterscharter.org/web/guest/about-the-charter, (accessed 20.08.2019).

[277] International Charter Space & Major Disasters, Oil Spills, https://disasterscharter.org/web/guest/disaster-types/-/article/oil-spills, (accessed 26.09.2019).

Oil spills occur when petroleum oil is released into the ocean following accidents, such as vessels crashing or damage and problems with oil platforms and drilling. They can have devastating effects

showed that the Charter has been frequently activated for weather-related disasters and only marginally for oil spills, but even so there are more than 60 registered cases of oil spills as can be seen on the website: "weather-related disasters such as floods, ocean and wind storms, landslides triggered by heavy rainfall or floods, wildfires, ice/snow hazard—representing 78% of all Charter activations—while solid earth-related hazards (e.g. earthquakes, volcanic eruptions) represent 17% of all Charter activations. Activations for oil spills and industrial accidents are marginal".[278]

2.5.5 *Illegal Fishing*

Illegal, Unreported and Unregulated (IUU) fishing means that vessels are not fulfilling their obligations to record and report catch or catch-related data. Illegal fishing is defined as "the targeting, catching, retaining on board, landing, sale or possession of fish which contravenes international, EU, UK or local fisheries law". Unreported fishing is defined as including "all landings and sales that do not get reported as either catches, landings, sales, or purchases". Unregulated fishing includes "activity which falls outside of any legislation yet remains legal or semi-legal". IUU has been the subject of different studies including by the European Parliament. Article 3 of IUU Regulation (EC) No 1005/2008 describes what is presumed to be involved in IUU fishing.[279] The EU tools against IUU fishing include: (i) the EU IUU vessel list; (ii) the list of non-cooperation third countries; (iii) emergency measures; (iv) port state

---

on the environment, affecting marine and coastal ecologies, and are not easy to remove. The ocean tide, currents and weather affect the movement of oil spills, making them unpredictable. Radar data is typically used to detect oil spills with satellites, but it can be very difficult to locate oil spills as looking for relatively small patches of oil in vast oceans is a challenge. Identifying oil is also difficult, as a dark patch on the ocean may be from the event in question, or a natural release of oil or different substance altogether. Once a particular oil spill that needs to be monitored is identified, however, satellites can play a useful role in noting the source, mapping the extent and tracking the direction the spill may take.

[278] International Charter "Space & Major Disasters", 2017 Annual Report, pp. 22, 2018, https://disasterscharter.org/documents/10180/14622/17th+Annual+Report.pdf, (accessed 20.08.2019).

[279] Council Regulation (EC) No 1005/2008 of 29 September 2008 establishing a Community system to prevent, deter and eliminate illegal, unreported and unregulated fishing, OJ 286, 29.10.2008, pp. 1–32, http://data.europa.eu/eli/reg/2008/1005/oj, (accessed 26.09.2019).

Article 3 of the IUU Regulation:

(a) fished without a valid license, authorization or permit issued by the flag State or the relevant coastal State; or

(b) not fulfilled its obligations to record and report catch or catch-related data, including data to be transmitted by satellite vessel monitoring system, or prior notices under Article 6; or

(c) fished in a closed area, during a closed season, without or after attainment of a quota or beyond a closed depth; or

(d) engaged in directed fishing for a stock which is subject to a moratorium or for which fishing is prohibited; or

(e) used prohibited or non-compliant fishing gear; or

(f) falsified or concealed its markings, identity or registration; or

(g) concealed, tampered with or disposed of evidence relating to an investigation; or

control; (v) catch certification scheme; (vi) the EU alert system; and (vii) mutual assistance.[280] Reports on illegal, unreported or unregulated fishing highlight that on average, one in five fish in global markets are caught by illegal fishing.[281]

The International Maritime Organization requires fishing vessels weighing 300 gross tones and over to carry Automatic Identification Systems (AIS), but for smaller vessels there is no such requirement. The problem is when vessels intentionally disable the AIS devices or report a series of false locations. Turning off the data to be transmitted by satellite vessel monitoring systems is a type of infringement. It could be reasonably assumed that vessels that are not transmitting location signals are taking part in illegal fishing. Illegal fishing is a problem for various reasons.[282] First, because catching more fish that can be sustainably harvested is a long-standing problem. Marine fish are a precious natural resource and their exploitation needs to be managed sustainably. Traceability is generally acknowledged as being a highly powerful tool in support of monitoring, control and enforcement in the fisheries sector. Second, because illegal fishing has consequences on the global economy as it exceeds USD20 billion annually in lost tax revenues, lost economic opportunities, and lost jobs.[283]

(h) obstructed the work of officials in the exercise of their duties in inspecting for compliance with the applicable conservation and management measures; or the work of observers in the exercise of their duties of observing compliance with the applicable Community rules; or

(i) taken on board, transshipped or landed undersized fish in contravention of the legislation in force; or

(j) transshipped or participated in joint fishing operations with, supported or re- supplied other fishing vessels identified as having engaged in IUU fishing under this Regulation, in particular those included in the Community IUU vessel list or in the IUU vessel list of a regional fisheries management organization; or

(k) carried out fishing activities in the area of a regional fisheries management organization in a manner inconsistent with or in contravention of the conservation and management measures of that organization and is flagged to a State not party to that organization, or not cooperating with that organization as established by that organization; or

(l) no nationality and is therefore a stateless vessel, in accordance with international law.

[280] Directorate-General for Internal Policies, Policy Department Structural and Cohesion Policies "Illegal, Unreported and Unregulated Fishing: Sanctions in the EU" Study 2014, http://www.europarl.europa.eu/RegData/etudes/STUD/2014/529069/IPOL_STU(2014)529069_EN.pdf, (accessed 5.08.2019).

[281] Ship Technology, Transparent transshipping: detecting illegal fishing with satellite data, 20 August 2018, https://www.ship-technology.com/features/global-fishing-watch/, (accessed 5.08.2019).

[282] Angling Trust, Stamping Out Illegal, Unreported and Unregulated fishing in UK, https://www.anglingtrust.net/page.asp?section=1167&sectionTitle=STAMPING+OUT+ILLEGAL%2C+UNREPORTED+%26+UNREGULATED+FISHING+IN+UK, (accessed 24.09.2019).

[283] Imagesat International, Using Satellite Imagery to Combat Illegal Fishing, 17 July 2017, https://www.maritime-executive.com/blog/using-satellite-imagery-to-combat-illegal-fishing, (accessed 24.09.2019).

Vessels that are not transmitting AIS signals can be detected during night-time surveillance.[284] Also, by accessing satellite data, governments and authorities can identify vessels and analyse if they are authorized to fish in their waters or in protected areas. The difference between illegal and legal fishing is the licence and the way the activity is performed. Satellite data can help in detecting which vessels are not complying with the law and environmental practices. For example, the Global Fishing Watch (GFW) is a non-profit organization that runs the GFW platform which uses data collected from space in order to map the movements of fishing vessels. Detecting illegal fishing and supporting transparent transhipping with satellite data is a business that could be further developed with the advent of small satellites and constellations in LEO to have uninterrupted global coverage. Artificial Intelligence (AI) and satellites could be used to combat illegal fishing.[285] An example for machine learning is OceanMind, which has found solutions to various challenges such as wide search areas.[286]

Ship detection using satellite remote sensing is vital for maritime surveillance. Among others, satellite imagery enables the monitoring of maritime traffic, illegal fishing and sea border activities. The AIS uses VHF radio frequencies to broadcast ship location, destination and identity. AIS is very effective for ships operating legally. But AIS can be switched off and in this case, satellite imagery is used to detect uncooperative ships. Synthetic Aperture Radar (SAR) imagery can very easily detect the radar signature of a ship and provide details such as ship dimensions, orientation and location. Unlike optical imagery, the SAR can be used in any conditions, including during night and cloudy skies. If combined, AIS and SAR could be a powerful tool for maritime surveillance, with SAR helping to monitor ships that may not want to be monitored.[287]

Illegal fishing ships use various means for disobeying the law. In a recent example, the Panama-flagged MV NIKA was intercepted by the Indonesian IUU fishing Task Force and captured by the Indonesian authorities. MV NIKA was suspected of illegal fishing operations and had changed its name seven times and switched flags seven times. So-called flag-hopping is a practice used by vessels wishing to cover their tracks. The movement of the ship could be tracked only with the help of satellite monitoring and GFW provided satellite imagery and tracking because the tracking

---

[284] Ship Technology, Transparent transshipping: detecting illegal fishing with satellite data, 20 August 2018, https://www.ship-technology.com/features/global-fishing-watch/, (accessed 24.09.2019).

[285] Thornton, A., How AI and satellites are used to combat illegal fishing, 6 June 2019, https://news.microsoft.com/on-the-issues/2019/06/06/ocean-mind-illegal-fishing/, (accessed 24.09.2019).

[286] *OceanMind's system currently tracks thousands of boats, with the capability of tracking millions, across the globe by gathering data from a wide range of sources, including collision-avoidance transponders aboard boats; radar images; satellite imagery; and cellphone signals. Analyzing these enormous datasets is beyond the capability of any one person. OceanMind has developed machine-learning algorithms that predict the type of fishing behavior based on vessel location, and flags suspicious and potentially illegal activity such as fishing too close to the shore.*

[287] Geocento, Ship detection from space, Satellite Imagery helps to monitor the seven seas, https://geocento.com/satellite-imagery-case-studies/satellite-imagery-can-help-on-maritime-surveillance/, (accessed 19.08.2019).

system required on all vessels of that size had been turned off by MV NIKA. As was noticed when in custody, despite being registered as a general cargo vessel, MV NIKA was found with fish processing facilities on board, which included "crab pots and other fishing gear ready to be used on deck, as well as herring for bait".[288] The CEO of GFW describes satellites as "gold-plated enforcement assets" and underlines that not many can afford to own such satellites but rather they should take the benefit of collaboration to use available technology.[289] It is reasonable to assess that small satellites could become the next generation of remote sensing satellites that would be even cheaper to use than their "gold-plated" counter-parts.

Several activity layers are available in the maps provided by the GFW, each of them containing very useful information collected by satellites. The first layer, described as the fishing effort activity layer, contains data about the vessel's identity, type, location, speed and direction. Data is collected via satellite and terrestrial receivers using AIS. Such data is analysed by applying an algorithm to determine apparent fishing activity, focusing on a fishing activity heat map and then detecting the anomalies.[290] Other activity layers refer to Indonesian Vessel Monitoring System (VMS); Peruvian VMS activity; vessel encounters and night light vessel detections. For light vessel detections, the satellite detects lights not obscured by clouds and could identify fleets that are not broadcasting AIS. However, the majority of lights detected at sea by night come from the commercial fishing vessels. Additional data is used to overlay with the activity layers, such as data about Exclusive Economic Zones (EEZ), Regional Fishery Management Organizations, High Seas Pockets and Marine Protected Areas.[291]

Another satellite system developed for detecting behaviour patterns of IUU fishing is the ImageSat International (ISI) developed by Kingsisher. This is a multi-sensor, multi-layer maritime intelligence system, combining satellite imagery to expose IUU fishing fleets.[292] Deep learning algorithms are used to identify vessels using satellite imagery and to correlate this with other sensor data such as satellite automatic identification, radar satellite imagery, electro-optical satellite imagery, VMS, coastal radar, open source intelligence and weather patterns Kingfisher detects non-cooperative vessels with AIS turned off. This system combines an AI system for greater effectiveness. The predictive behaviour of the ships is essential for the coast guard to intercept the vessel. This system for IUU fishing could be as useful also for vessel

[288] Bladen, S., The capture of the MV Nika: A case of illicit fishing and a showcase for how to beat it, 23 July 2019, https://globalfishingwatch.org/vms-transparency/the-capture-of-the-mv-nika-a-case-of-illicit-fishing-and-a-showcase-for-how-to-beat-it/, (accessed 16.08.2019).

[289] Ibid. *supra* note Bladen, S., The capture of the MV Nika: A case of illicit fishing and a showcase for how to beat it, 23 July 2019.

[290] Global Fishing Watch, The Global Fishing Watch Map, https://globalfishingwatch.org/our-map/, (accessed 24.09.2019).

[291] Ibid. *supra* note Global Fishing Watch, The Global Fishing Watch Map, https://globalfishingwatch.org/our-map/, (accessed 24.09.2019).

[292] Ibid. *supra* note Imagesat International, Using Satellite Imagery to Combat Illegal Fishing, 17 July 2017, https://www.maritime-executive.com/blog/using-satellite-imagery-to-combat-illegal-fishing, (accessed 24.09.2019).

movements related to piracy and terrorism, drug smuggling, human trafficking and illegal immigration.[293]

2.5.6 *Amazon: Rainforest Exploitation and Wildfires*

The Amazon River basin is home to the largest rainforest in the world with an impressive concentration of flora and fauna, covering about 2.67 million square miles (6.9 million square kilometres) in seven countries,[294] responsible for about 40% of the world's rainforest absorption of carbon dioxide.[295] The Amazon rainforest is often referred to as "the planet's lungs" because it provides almost 20% of the world's oxygen and is a vital carbon store that slows down the pace of global warming.[296] The rainforest of the Amazon Basin absorbs carbon dioxide (CO_2) during the process of photosynthesis and the Amazon trees account for a quarter of the carbon dioxide absorbed.

Recent studies have revealed that the Amazon rainforest is taking up a third less carbon than a decade ago.[297] Also, it is alarming that the environment of the Amazon Basin is changing due to extreme conditions such as burning fires or deforestation and the Amazon rainforest could produce more carbon than absorbs. The intense drought from 2005 created temporary conditions for the forest to become a net producer of greenhouse gases. In fact, the drought sharply reversed the growth of trees and the die-off rate doubled to 2%. This was described as a stark reminder of global warming consequences.[298] Carlos Peres, Biologist at the University of East Anglia, has noted that high moisture protects the Amazonian forest from fires but this changed during severe droughts in 1997–1998, 2005 and 2010. Drought combined with a multitude of fires across northern Brazil could arbitrarily change the environment. To better understand the whole complexity, in addition to deforestation and wildfires, the

---

[293] Ibid. *supra* note Imagesat International, Using Satellite Imagery to Combat Illegal Fishing, 17 July 2017, https://www.maritime-executive.com/blog/using-satellite-imagery-to-combat-illegal-fishing, (accessed 24.09.2019).

[294] Brazil, Bolivia, Colombia, Ecuador, Guyana, Peru, Suriname and Venezuela, with 60% of the rainforest area falling under the Brazilian territory.

[295] *Humans worldwide are estimated emit about 32 billion tons of carbon dioxide each year but just less than half of this, about 15 billion tons, remains in the atmosphere. The rest is absorbed by natural carbon sinks in the ocean and on land. Scientists have calculated that the world's tropical forests collectively absorb about 4.8 billion tons of carbon dioxide every year, with the Amazon being the single biggest rainforest sink. Amazonia alone is estimated to store about 100 billion tons of carbon locked up in its trees.*

[296] Connor, S., Revenge of the rainforest, 06 March 2019, https://www.independent.co.uk/environment/climate-change/revenge-of-the-rainforest-1638524.html, (accessed 11.08.2019).

[297] Brienen, R.J.W, Philips, O.L., et al. Long-term decline of the Amazon carbon sink, Nature, 19 March 2015, https://doi.org/10.1038/nature14283, (accessed 24.09.2019).

[298] Ibid. *supra* note Connor, S., Revenge of the rainforest, 06 March 2019, https://www.independent.co.uk/environment/climate-change/revenge-of-the-rainforest-1638524.html, (accessed 11.08.2019).

situation in the Amazon is becoming even more alarming because trees are dying faster. The mortality rate has increased "due to the increase in temperature, more prolonged and worse droughts, stronger winds, more extensive fires, more vines, and even the abundance of carbon dioxide in the atmosphere".[299]

The tropical rainforests of the Amazon Basin face the threat of deforestation and satellite surveys indicate that about 5,800 square miles of the Amazon rainforest are destroyed each year. Deforestation is happening largely for reasons such as: (i) farming; (ii) logging; (iii) mining; (iv) roads; hydroelectric power (HEP); population growth and infrastructure. Observations from satellites have concluded that the state of Rondônia in western Brazil has become one of the most deforested parts of the Amazon.[300] Deforestation of the Brazilian Amazon has negative consequences as the steady erosion of tree cover weakens the role of the rainforest in stabilizing the global climate.[301]

Real time deforestation detection through satellite images could be among the game changing solutions to save the Amazon rainforest. In a 2015 study it was underlined that efficient technology would require using a sensor with high temporal or spatial resolution that increases the likelihood of obtaining images without the incidents of clouds. The scientific proposal included the development of a tool for detection in real time (daily) using the sensor Moderate Resolution Imaging Spectroradiometer (MODIS) of the TERRA satellite and ANN, which is an AI technique capable of detecting changing patterns.[302] The EO program led by NASA launched the TERRA satellite in 1999 containing the MODIS sensor designed to observe the Earth's surface, oceans and the atmosphere in the visible and infrared electromagnetic spectrum.[303]

Deforestation forecasting is a difficult task since many factors can influence it such as socio-economic, environmental and cultural change. However, one of the variables can be analysed from historical data provided by satellite images over a longer period of time in order to see the areas most likely challenged by deforestation. In this regard, using satellite images could monitor the dynamics of deforestation

---

[299] Amorim, C., Trees in tropical rainforest are dying faster, 01.06.2018, https://ipam.org.br/trees-in-tropical-rainforests-are-dying-faster/, (accessed 25.08.2019).

[300] NASA Earth Observatory, *In the past three decades, clearing and degradation of the state's forests have been rapid: 4,200 square kilometers cleared by 1978; 30,000 by 1988; and 53,300 by 1998. By 2003, an estimated 67,764 square kilometers of rainforest—an area larger than the state of West Virginia—had been cleared.* https://earthobservatory.nasa.gov/world-of-change/Deforestation, (accessed 24.09.2019).

[301] Watts, J., Amazon deforestation accelerating towards unrecoverable "tipping point", The Guardian, 25 July 2019, https://www.theguardian.com/world/2019/jul/25/amazonian-rainforest-near-unrecoverable-tipping-point, (accessed 17.08.2019).

[302] Kehl, T.N., Todt, V., Veronez, M.R., Cazella, S.C., Real time deforestation detection using ANN and Satellite images—The Amazon Rainforest study case, pp. 3, https://doi.org/10.1007/978-3-319-15741-2, Springer, Cham, 2015.

[303] Ibid. *supra* note Kehl, T.N., Todt, V., Veronez, M.R., Cazella, S.C., Real time deforestation detection using ANN and Satellite images—The Amazon Rainforest study case, pp. 9.

and AI techniques in the classification of orbital data could obtain better results.[304] There are many examples of sensors monitoring the Amazon including LANDSAT and COPERNICUS SENTINEL.

In August 2019, satellite images revealed the Brazilian Amazon was burning at a record rate of fires. These are the worst events in recent history showing burning fires destroying the Amazon and tearing across Brazil at a hitherto unseen pace. More than 75,000 fires were detected in 2019 alone according to Brazil's space research centre, Instituto Nacional de Pesquisas Espaciaisn (INPE). The INPE, using satellite images, identified 9,500 new forest fires in August. Satellite imagery from NASA and the National Oceanic and Atmospheric Administration (NOAA) provided satellite images that underlined that Roraima was coated in dark smoke.[305] Observing from space the wildfires in the Amazon rainforest, revealed the areas where the smoke spread across South America. NOAA/NASA provided satellite images from the SUOMI NPP satellite using Visible Infrared Imaging Radiometer Suite (VIIRS) instruments showing smoke above the Brazilian states of Amazons, Para, Mato Grosso and Rôndonia. Also, satellite imagery from the MODIS on NASA's Aqua satellite provided scientists with the level of progression of the wildfires. Other satellite imagery was provided by NASA's EO System Data and Information System (EOSDIS), a worldview application that allows tracking fires around the world using NASA satellites.[306] The strong winds carried the intense smoke more than 2,700 km to São Paulo.

The PLANET's constellation of satellites was able to provide high-resolution images from their "near-daily overflights" over areas with fires devastating the Amazon rainforest. It was underlined that with the current satellites in orbit, a "near real-time view of what's unfolding in the Amazon" can be provided. The data is used to assess the impact of the Amazon fires on carbon emissions. The Amazon rainforest stores carbon dioxide in its trees and if burned up it could have a devastating impact on climate change.[307]

> If you took all of the carbon stored in every tropical forest on Earth and burned it up, you would emit about five times the carbon dioxide into the atmosphere that is already there. The Amazon rainforest represents about half of this forest carbon to give you an idea of how serious this current situation is and the kind of impact it will have on climate change.

---

[304] Ibid. *supra* note Kehl, T.N., Todt, V., Veronez, M.R., Cazella, S.C., Real time deforestation detection using ANN and Satellite images—The Amazon Rainforest study case, pp. 18.

[305] Langton, K., Amazon fires satellite image: Devastating Brazil rainforest fire burning at "record rate", 28 August 2019, https://www.express.co.uk/news/world/1168303/amazon-fires-satellite-image-amazon-rainforest-wildfires-inpe-brazil-space-agency, (accessed 24.09.2019).

[306] Gohd, C., As Fires Devastate the Amazon Rainforest, NASA Satellites Capture Grim Images, 23 August 2019, https://www.space.com/amazon-rainforest-fires-nasa-satellite-images.html, (accessed 25.08.2019).

[307] Butler, R.A., Satellite images from Planet reveal devastating Amazon fires in near real-time, 22 August 2019, https://news.mongabay.com/2019/08/satellite-images-from-planet-reveal-devastating-amazon-fires-in-near-real-time/, (accessed 25.08.2019).

The Amazon Environmental Research Institute (IPAM) is a Brazilian research group. In a recent study IPAM used data sources derived from satellite imagery to compare fire seasons. The satellite imagery captured the dynamics of fire activity and its relationship with deforestation and maximum dry length meaning the maximum of consecutive days with precipitation below 1 mm. The dataset used for fire counts originated from the AQUA satellite, which is considered a reference satellite by the INPE, and the dataset used for determining the maximum dry spell lengths originated from the Climate Hazards Group InfraRed Precipitation with Stations (CHIRPS). The results of this analysis showed that deforestation was the cause of most of the fire activity in Amazonia in 2019, with climatic conditions such as severe drought having a very low explanatory power:

> The top ten Amazon municipalities in fire counts were also the ones deforesting the most (…). Considering all municipalities in Amazonia, those 10 were responsible for 37% of the fire counts and 43% of the accumulated deforestation up to July in 2019 (…). This concentration of fires in municipalities experiencing high deforestation rates in a non-drought year strongly indicates that the fires of 2019 have been intentional and mostly for clearing recently deforested areas.[308]

Therefore, IPAM raised the problem of Amazon wildfires, indicating deforestation as the source of wildfires rather than attributing the fires to the drought and the dry season. IPAM rejects the intense drought as the source of fire because, despite being the dry season, analysis has shown that moisture levels in the Amazon are above average compared to the last three years. IPAM further stated that fire is commonly used to clear the terrain after forests are felled and as an example mentioned a less severe drought such as the 2016 El Nino effect, when the growing fires were out of control.[309] However, droughts and strong winds create the conditions for fire to spread widely, increasing chances of large scale disturbances such as in August 2019.

Deforestation and wildfires could easily get out of control and the situation could be better assessed from space using satellite imagery. Smoke creates problems in the regions which are close or at even 3,000 km distance such was the case of Sao Paulo. Air pollution from the smoke generated by Amazon fires could negatively impact the environment, social and economic activity and it is essential to continue to use satellite imagery and monitor the impact of wildfires while using this information to also prevent and stop deforestation and wildfires.

---

[308] Silveiro, D., Silva, S., Alencar, A., Moutinho, P., Amazon on fire: Technical note from the Amazon environmental Research Institute—IPAM, 2019, https://ipam.org.br/wp-content/uploads/2019/08/NT-Fogo-Amazônia-2019_English_v2.pdf, (accessed 25.08.2019).

[309] IPAM, Deforestation, not the dry season, responsible for Amazon burning in 2019, 23 August 2019, https://ipam.org.br/deforestation-not-the-dry-season-responsible-for-amazon-burning-in-2019/, (accessed 25.08.2019).

> There are no natural fires in the Amazon. There are, however, people who light fires, which can worsen and turn into wildfires during the dry season," explains IPAM Science Director Ane Alencar, one of the authors of the memo. "Even in a less severe drought than in 2016, when we suffered from a very strong El Niño effect, the risk of fires growing out of control is high.

2.5.7 *Practice of Monitoring Compliance with International Human Rights and Environmental Agreements*

Compliance can be defined as 'fulfilment by the contracting parties of their obligations under an agreement and any amendments to the agreement thereof.'[310] Satellite data is a key source for collecting information that is useful in the implementation of international human rights and environmental agreements. Through its various judgments the ICJ has upheld the importance of satellite-based data and its admissibility. Monitoring compliance with international human rights and environmental law is possible through satellite data, as it can be utilized to gain a better understanding of the overall situation from a neutral perspective. In recent times, monitoring compliance has become an issue of great concern. Satellite data enables the rapid and unhindered collection of data relevant and factual data along with imageries to prove violations on any location on Earth.[311]

Remote sensing plays a vital role in enhancing the ability to monitor treaty compliance. Remote sensing assists governments comply with their obligations and improve performance under various human rights and environmental agreements. The availability of satellite data could also support private actors in conforming with regulations set up at the domestic level.[312] Technological developments have greatly enhanced the accuracy of the data collected and have also resulted in satellite data being extremely useful in implementing international law. The ability of such data to provide factual, relevant and up-to-date information has made it easier to monitor compliance of obligations under international agreements.[313]

The aim of environmental agreements is to protect human health and the environment as a whole. States parties need to ensure they have the technical capabilities to fulfil their obligations according to these agreements. However, evaluating compliance with these obligations is not an easy task. Also, monitoring compliance plays a pivotal role in ensuring that states are bound by these agreements. Monitoring compliance enables international organizations to verify the correctness of the declarations made by States parties to environmental agreements and provides detailed information for on-site inspections. Considering the consequences of human rights violations and damage being caused to the environment, the UN along with governments and

---

310 UNEP United Nations Environment Programme, Manual on Compliance with and Enforcement of Multilateral Environmental Agreements, pp. 59, 2006, https://wedocs.unep.org/bitstream/handle/20.500.11822/7458/-Manual%20on%20Compliance%20with%20and%20Enforcement%20of%20Multilateral%20Environmental%20Agreements-2006743.pdf?sequence=3&isAllowed=y, (accessed 24.09.2019).

311 Annette Froehlich, "*Space Related Data: From Justice to Development*," IAC 2011 (IAC-11.E7.3.1) at 5.

312 Yahya A. Dehqanzada and Ann M. Florini, "*Secrets for Sale: How Commercial Satellite Imagery Will Change the World*," Carnegie Endowment for International Peace (2000) at 11.

313 Ridderhof, R., Satellite Data in International Law, Peace Palace Library, https://www.peacepalacelibrary.nl/2017/08/satellite-data-in-international-law/ (accessed 05.06.2019).

NGOs is required to monitor compliance.[314] The Office of the High Commission for Human Rights utilizes remote sensing as a tool to improve deterrence and detection of human rights violations. It helps in the active collection, verification and use of information.[315]

Satellite data or remote sensing is not only important as substantive evidence but also helps complement ground-based compliance mechanisms as it can provide accurate, objective and comparable data without infringing upon national sovereignty.[316] Its ability to gather information covering vast areas makes it a more efficient and reliable source of evidence.[317] It has a distinct advantage over other conventional measurements as:

- it allows collection and comparison of data over different geographical locations,
- it allows for remote measurements,
- it provides raw satellite data that can be reprocessed and made commonly accessible,
- it also allows for continuous measurement of data, which helps provide a better comparative analysis.

While satellite data do not provide all the required information that certain treaties require, they are necessary to develop and implement Multilateral Environmental Agreements (MEAs). One such ambitious agreement to utilize satellite data has been the European Commission's Global Monitoring for Environment and Security (GMES).[318] GMES strives to utilize remote sensing for three main reasons:

- Environmental Treaties,
- Natural Disasters and,
- Humanitarian Aid.

Another instrument that specifies the use of remote sensing is the Johannesburg Declaration that refers to EO as a crucial source of information and a key decision-making tool for better management of water resources, natural disasters and climate monitoring. With the development of technology, remote sensing is also being used as evidence in international courts. Some of the cases where satellite data have been accepted as evidence are the Singapore Song San case concerning oil pollution

---

[314]Edwards, S., Koettl, C., Looking to the Sky: Monitoring Human Rights through Remote Sensing, Harvard International Review, 32 (4), pp. 66-71, 2011.

[315]Marx, A., Goward, S.N., Remote Sensing in Human Rights and International Humanitarian Law Monitoring: Concepts and Methods, The Geographical Review, 103(1), pp. 100, 2013.

[316]Uhlir, P.F., From Spacecraft to Statecraft: The Role of Earth Observation Satellites in the Development of International Environmental Protection Agreements, GIS Law, Vol. 2, No. 3, 1995.

[317]Sherninin, A., Giri, C., Remote Sensing in Support of Multilateral Environmental Agreements: What Have We Learned from Pilot Applications? Open Meeting of the Human Dimensions of Global Environmental Change Research Community, Rio de Janerio, 6-8 October 2001, https://www.researchgate.net/publication/228792200_Remote_Sensing_in_Support_of_Multilateral_Environmental_Agreements_What_Have_We_Learned_from_Pilot_Applications, (accessed 5.06.2019).

[318]Ibid. *supra* note Sherninin, A., Giri, C., Remote Sensing in Support of Multilateral Environmental Agreements: What Have We Learned from Pilot Applications?

in the Strait of Malacca, the Qatar/Bahrain Maritime and Territorial Delimitation Case, the Nigeria/Cameroon's dispute concerning land and maritime boundaries, and Kasikili/Sedudu Island dispute between Botswana/Namibia.

Monitoring compliance is now a matter of great concern as agreements undergo numerous reviews and modifications. This has led to enforcement officials facing challenges that hamper their ability to ensure effective and efficient enforcement.[319] Remote sensing to monitor compliance is currently more theoretical than applied as developments are more technology—oriented and lack legal co-operation.

Though satellite data play a key role, it is still to be investigated and utilised by several treaties. It requires the treaties to set up a competent authority and provide a mandate for verification. Use of satellite data as a monitoring mechanism must be specifically mentioned in the treaties and agreements to make it a proper source for monitoring compliance. And considering the wide range of states and their scientific and technical capabilities, there is a need to provide proper training and consultancy to utilize satellite data. As a result, monitoring compliance depends from agreement to agreement. There are no general framework rules for monitoring compliance and authorities vary.[320]

2.5.8 *Importance of Remote Sensing for Monitoring Compliance*

While discussing the importance of remote sensing to monitor compliance it is important to acknowledge that remote sensing is in itself not sufficient as an independent source of information. Monitoring compliance requires satellite data to be utilized with on-site inspections and personal inquiries given that not all phenomena are visible and recordable by remote sensing. Despite certain shortcomings, remote sensing is an efficient tool available to monitor compliance under international law.

Monitoring compliance of international human rights and environmental agreements is important to build the confidence of states to meet their respective obligations. Remote sensing can be used to detect non-compliance, which helps in deterring parties that are non-compliant or unwilling to fulfil their obligations under these agreements.[321] Satellite data can help parties reassure their commitment to their obligations under these agreements while demonstrating that any violations or non-compliance will be detected. It can also help in the peaceful settlement of disputes. It is one of the mechanisms of monitoring compliance that affects state sovereignty

[319] Abbot, C., Enforcing Pollution Control Regulation, Strengthening Sanctions and Improving Deterrence, Hart Publishing, Oxford, pp. 1, 2009.

[320] Hettling, J.K., The Use of Remote Sensing Satellites for Verification in International Law, Space Policy 19, pp. 33, 2003.

[321] Findlay, Introduction: The Salience and Future of Verification, Verification Yearbook, pp. 16, 2000.

the least, while simultaneously adhering to the principle of non-interference under public international law.[322]

Remote sensing plays an important role in monitoring compliance because data collection from space is permitted under international law, and any interference with such an act is contrary to international law. It eases compliance monitoring mechanisms as they do not specifically need permission from the state being sensed. However, as the UNGA Principles on Remote Sensing are regarded as resolutions, they do not have a binding nature, which has led to a lack of acceptance of satellite data.[323] It is available for use not only to monitor compliance by regulatory authorities but by everyone. This creates a unique way of monitoring compliance, the involvement of NGOs in various circumstances has helped utilize satellite data to ensure compliance.[324] NGOs can utilize the data collected to shape public opinion and to create awareness, lobby governments and help monitor industry compliance with obligations.[325]

However, not all agreements have a compliance mechanism as it is believed that these can act as a deterrent for parties to ratify the treaty. There are also opinions that the presence of a compliance mechanism helps in confidence-building of the objectives of the agreement. It helps improve the effective and fair application of the agreement's guidelines. The presence of compliance procedures is believed to enhance the credibility of agreements and their implementation by states.[326]

Satellite data has played an important role as evidence in various cases before national and international courts and dispute settlement agencies. Despite its use, whether satellite data can be regarded as a proof is still debatable. Opinions are expressed that, as there is no proper definition of 'proof' under international law, interpretation under national jurisdictions has to be utilised. In most of the disputes, satellite data has been utilised as an inquiry tool to confirm or deny, rather than as a system of proof itself.[327] Regarding monitoring compliance, the procedure of remote sensing should be further developed to become a mature capability to ensure and inspect Party's compliance with their obligations under various agreements.

---

[322]Bothe, Verification of Facts, Encyclopedia of Public International Law IV, pp. 1268, 2002.

[323]Ibid. *supra* note Hettling, J.K., The Use of Remote Sensing Satellites for Verification in International Law, Space Policy 19, pp. 37.

[324]Zaelke, D., Kaniaru, D., Kružiková, E., Making Law Work: Environmental Compliance and Sustainable Development, Vol. 1 and 2, Cameron May, London, 2005.

[325]Purdy, R., Using Earth Observation Technologies for Better Regulatory Compliance and Enforcement of Environmental Law, Journal of Environmental law 22:1, pp. 64, 2010.

[326]Mullins, F., Kyoto Mechanism, Monitoring and Compliance—From Kyoto to The Hague, OECD Environment Directorate and International Energy Agency, COM/ENV/EPOC/IEA/SLT 9, pp. 36, 2001.

[327]UNECE, Environmental Monitoring, https://www.unece.org/environmental-policy/environmental-monitoring-and-assessment/areas-of-work/environmental-monitoring.html, (accessed 06.07.2019).

2.5.9 Remote Sensing and Human Rights Monitoring

Remote sensing is one of the tools that can help detect a human rights violation as acts such as the destruction of civilian property not justified by military necessity and indirect violations such as forcible transfer of people can be detected from space. It is also used in activities such as 'human-rights mapping,' where satellite images are used by UN peacekeeping missions and NGOs to conduct systematic human-rights monitoring campaigns and observation; and can be used as evidence in courts. Remote sensing plays a very prominent role as it allows organizations to identify violations they are attempting to detect. The organization monitoring compliance can even select a remotely sensed phenomenon that is usually associated with a human rights violation and then choose sensors to detect and analyse the data.[328]

This systematic collection of data allows the organizations to run human rights monitoring campaigns. One such prominent campaign is 'The AAAS Scientific Responsibility, Human Rights and Law Program.' This program aims to determine the human rights application of remote sensing and increase understanding among the courts on the use of remote sensing technology. The program focuses on pursuing the use of satellite data in human rights litigation at international and regional levels. The AAAS program has contributed greatly to the use of satellite data for monitoring compliance with human rights laws by training judges, investigators and prosecutors in remote sensing analysis.[329]

Satellite data, mapping and imaging help produce compelling corroborative evidence to eyewitness reports. High-resolution mapping can help detect small phenomena that are associated with human rights violations. It can help identify damages to infrastructure, mass graves, resource degradation, and track mass population movements. Monitoring compliance of human rights agreements can focus on incidents and events that are relevant to certain regions or populations that are potentially affected.[330]

Satellite data help human rights campaigning in reducing the time between eyewitness reports and imaging to produce corroborative evidence. It has the potential to provide relevant data in situations where fieldwork is not possible, desired or affordable. Remote sensing is particularly important in circumstances where states or their authorities refuse access to the region.[331] It enables continuous collection of data

[328] Ibid. *supra* note Marx, A., Goward, S.N., Remote Sensing in Human Rights and International Humanitarian Law Monitoring: Concepts and Methods, The Geographical Review, 103(1), pp. 100, 2013.

[329] AAAS, SRHRL Past Projects: Remote Sensing for Human Rights, https://www.aaas.org/programs/scientific-responsibility-human-rights-law/past-projects/remote-sensing-human-rights, (accessed 10.07.2019).

[330] Pierro, R., Satellite Imagery for Human Rights Monitoring, The Engine Room Library, https://library.theengineroom.org/satellite-imagery-human-rights/#introduction, (accessed 09.07.2019).

[331] Schneiderbauer, S., Monitoring Multilateral Humanitarian Agreements, In: Wirkus, L., Vollmer, R., Monitoring Environment and Security—Integrating Concepts and Enhancing Methodologies, BICC Seminar Documentation, Brief 37, pp. 47, July 2008, https://www.researchgate.net/publication/283321139_Monitoring_Environment_and_Security_-_Integrating_concepts_and_enhancing_methodologies, (accessed 8.07.2019).

to facilitate detection of changes at certain periods. Remote sensing plays a prominent role in monitoring compliance of humanitarian and environmental agreements, as without proper compliance there is no deterrence possible and obligations under environmental law would only stand as symbolic rather than efficient.[332]

Satellite imaging has helped in various circumstances to prove human rights violation. One such incident was the Russian Occupation of Tskhinvali, a part of the Republic of Georgia, where imaging enabled the American Association for the Advancement of Science and Amnesty International to prove the destruction of Georgian homes by Russian forces.[333]

2.5.10 Despite Its Use in Monitoring Compliance, There Are Several Concerns About Satellite Data

- **Weather dependence**: the majority of EO satellites that are capable of capturing data relevant to human rights issues are usable only under cloud-free conditions.
- **Shutter control**: though remote sensing principles allow the collection of data, there is always a possibility of governments prohibiting agencies and companies from distributing images that could prove human rights violation. One such instance was during the war in Afghanistan when the U.S. prohibited the sale of satellite data concerning Afghanistan.
- **Legal uncertainty**: due to legal issues concerning remote sensing without consent, states that are obliged to comply with treaty obligations may refuse to accept the investigation reports.

Examples where remote sensing has been used to monitor compliance with humanitarian agreements:

- **Kosovo (1997)**—the Kosovo conflict saw the Kosovo Liberation Army fight for independence from Serbia. This conflict saw the killing of several thousand Kosovars by the Serbian Army and nearly 150,000 Kosovo Albanians becoming homeless. In response, NATO carried out military operations in 1999 that were justified by aerial photographs that proved human rights violations.
- **Zimbabwe (2005)**—Operation Murambatsvina, which was carried out by the President Mugabe run Zimbabwean Government saw an estimated 700,000 people lose their homes and livelihood. Satellite images showed the extent of damage done by the operation.[334]

[332]Macrory, R., Regulation, Enforcement and Governance in Environmental Law, Cameron May, London, pp. 771, 2008.

[333]Ibid. *supra* note Marx, A., Goward, S.N., Remote Sensing in Human Rights and International Humanitarian Law Monitoring: Concepts and Methods, The Geographical Review, 103 (1).

[334]Ibid. *supra* note Schneiderbauer, S., Monitoring Multilateral Humanitarian Agreements, in: Wirkus, L., Vollmer, R., Monitoring Environment and Security—Integrating Concepts and Enhancing Methodologies, BICC Seminar Documentation, Brief 37, pp. 50, July 2008.

- **Sudan (2007)**—during the conflict between different ethnic and tribal groups, the Sudanese government was accused of multiple human rights violations. AI utilized satellite data to analyse the situation to prove demolition of villages and the rapid growth of refugee camps.

2.5.11 *Use of Remote Sensing to Monitor the Application of Multilateral Environmental Agreements*

Most Multilateral Environmental Agreements (MEAs) oblige States directly or indirectly to implement certain mechanisms to monitor and assess the environment regularly. There is a growing number of MEAs who now using remote sensing and EO data to ensure compliance. Though current satellites are not designed to meet all the compliance requirements they do help generate pertinent information for developing and implementing agreements. MEAs are usually contracted between states to address environmental issues and utilize various tools to accomplish policy objectives. Remote sensing is one such tool that has contributed to drafting relevant and improved policy instruments.[335]

The use of remote sensing for the application of MEAs can range from identification of new environmental issues, monitoring and assessment of the problem, and verification of compliance and enforcement. However, there is no international binding regime to specifically address the issue of monitoring compliance through remote sensing.[336] Remote sensing has the potential to refine and develop MEAs. It can be used to understand and evaluate the effectiveness of environmental agreements, while also assisting states in better implementation of their obligations.[337] MEAs lay down objectives to be achieved and actions to ensure a better environment. Satellite data and its ability to convey environmental changes in a compelling way is also necessary to create awareness and raise political support to strengthen MEAs. As a result, environmental agreements usually have a softer compliance monitoring mechanism, with the main emphasis on cooperation rather than confrontation.[338]

---

335 Peter, N., The Use of Remote Sensing to Support the Application of Multilateral Environmental Agreements, Space Policy, 20, pp. 189, 2004.

336 Ibid. *supra* note Peter, N., The Use of Remote Sensing to Support the Application of Multilateral Environmental Agreements, Space Policy, 20, pp. 190, 2004.

337 Arino, O., Fernnádez-Prieto, D., Paganini, M., Volden, E., and Seifert, F.M., Earth Observation Supporting Multilateral Environmental Agreements: The Initiatives of the European Space Agency, Proc. IEEE. International Geoscience and Remote Sensing Symposium (IGARSS'03), 2003.

338 Tenner, Verification Mechanisms in International Environmental Agreements, VERTIC Briefing Paper 99/2, pp. 1, 1999.

2.5.11.1 Remote Sensing and Its Uses in Various Multilateral Environmental Agreements

Many MEAs require direct or indirect monitoring. Different agreements have different objectives and implementation practices. Satellite data has the potential to assist in achieving these objectives. It can also be an efficient management system to address various environmental issues and contribute to improving commitment to the agreement.[339] There are different purposes for which satellite data can be utilized with regard to environmental agreements:[340]

- **Targeted Enforcement**

Remote sensing can be used as targeted enforcement strategy to monitor legislation. One such use can be seen in the Australian legislation, where to curb illegal deforestation for farming, Australian states utilize satellite data to monitor tree clearing as per the policy guidelines. Satellite data provides an efficient way of surveillance by monitoring land clearance for measurement and comparison. Anther prominent use of targeted enforcement can be seen in the EU agricultural sector, where the EU allows the member states to utilise satellite data to monitor environmental conditions relating to farm subsidy payments.[341] The satellite monitoring system has also been used in the enforcement of fisheries, where it is mandatory for certain fishing vessels to install satellite monitoring equipment to track the movement of the vessels. This helps monitor fishing in authorized areas.[342]

- **Monitoring High-Risk Offenders**

Remote sensing can also be utilized by regulators to monitor compliance in cases of prosecution. It is used to monitor high-risk sites to ensure that decisions imposed in a successful prosecution are being adhered to. It is especially useful where the area to be monitored is vast and not easily accessible.

- **Historical Evidence**

Collection of satellite data and utilizing data archives can be of great importance as historical evidence. It has the potential to provide an impartial view of any location or

---

[339] Ibid. *supra* note Arino, O., Fernnádez-Prieto, D., Paganini, M., Volden, E., and Seifert, F.M., Earth Observation Supporting Multilateral Environmental Agreements: The Initiatives of the European Space Agency, Proc. IEEE. International Geoscience and Remote Sensing Symposium (IGARSS'03), 2003.

[340] Ibid. *supra* note Purdy, R., Using Earth Observation Technologies for Better Regulatory Compliance and Enforcement of Environmental Law, Journal of Environmental law 22:1, pp. 64, 2010.

[341] Council Regulation (EC) No. 1782/2003 of 29 September 2003 establishing common rules for direct support schemes under the common agricultural policy and establishing certain support schemes for farmers, *OJ L 270, 21.10.2003, pp. 1–69 (ES, DA, DE, EL, EN, FR, IT, NL, PT, FI, SV).*

[342] Council Regulation (EC) 2244/2003 of 18 December 2003 laying down detailed provisions regarding satellite-based Vessel Monitoring Systems, *OJ L 333, 20.12.2003, pp. 17–27 (ES, DA, DE, EL, EN, FR, IT, NL, PT, FI, SV).*

data during any point of time since data collection started. It helps provide an impartial analysis of the timing of the activity. Satellite data has been used as evidence in many important dispute settlements before the International Court of Justice.[343]

2.5.11.2 Remote Sensing and Agreements

Though linking remote sensing to policy is not straightforward. Some of the important agreements that utilize satellite data are as follows:

- **Antarctic Treaty System, 1959**

The Treaty mandates that Antarctica be used for peaceful purposes only and scientific investigation while encouraging cooperation between scientific personnel. CCAMLR[344] mandates sustainable harvesting of resources, while the Convention for the Conservation of Antarctic Seals mandates protection of seals in the region. The treaty bans mining activities and requires a compulsory environmental impact assessment before beginning activities.[345]

The treaty does not have an agency to monitor or enforce rules. However, the Committee on Environmental Protection (CEP) is responsible for evaluating environmental impact assessments of activities. As a result, monitoring compliance is the responsibility of the parties. The treaty allows for on-site inspection and aerial surveillance of stations and activities. As a result, satellite data can be utilized to monitor the activities of the parties. Though some satellite data concerning the Antarctic, such as surface temperatures, ice boundaries and thickness, snowmelt quantities, volcanic activities, sea surface, and temperature, may not be useful for monitoring compliance under the Treaty, it can be useful for other treaties such as the Kyoto Protocol. Though there is no mechanism under the Treaty to monitor compliance, scientific monitoring through remote sensing can help detect oil spills from ships[346] and station activities, smoke and aerosol plumes and illegal mining activities.

---

[343] Satellite data has been used as evidence in various cases including:

- Territorial dispute between Qatar and Bahrain—(1995) ICJ Report 6,
- Land and Maritime Boundary dispute between Cameroon and Nigeria—(2002) ICJ Report 30,3
- Kasikili/Sedudu Island dispute between Bostwana and Namibia—(1999) ICJ Report 1045.

[344] Convention on the Conservation of Antarctic Marine Living Resources, 1980.

[345] Joyner, C., Governing the Frozen Commons: The Antarctic Regime and Environmental Protection, University of South Carolina Press, pp. 353, 1998.

[346] There have been multiple instances of oil spillage in the Antarctic which included Argentine Ship Bahia Paraiso spilling 250,000 gallons of diesel in the Antarctic Peninsula, British ship, HMS Endurance spilling oil into the Esperanza Bay after hitting an iceberg, Peruvian vessel BIC Humboldt

- **Ramsar Convention on Wetlands, 1971**

The Convention obligates the parties to list wetlands sites of importance and to take necessary actions to establish reserves and increase waterfowl population. It also requires the implementation of policies and directs the parties every third year to submit national reports detailing implementation activities. The Ramsar Bureau set up under the Convention monitors compliance and verifies data submitted under the national reports.[347] Satellite data is useful for the Convention as states can map tropical forests and wetlands, which can then be combined with a field survey to prepare a biodiversity map. It also helps study and map wetland boundaries.

- **International Convention for the Prevention of Pollution from Ships (MARPOL), 1973**

This convention aims to eliminate pollution of the sea and oceans due to oil, chemical and harmful discharges from the ships. Remote sensing can help measure oil slicks and the release of harmful chemicals from ships into the oceans. MARPOL is one of the only agreements that recognize remote sensing as a potential compliance monitoring mechanism. The convention regulates the disposal of waste generated by vessels and seeks to reduce oil entering into the ocean. Based on the size and construction of the vessels, the convention specifies design and equipment standards. Vessels are required to monitor discharge and use methods to retain if necessary.[348] The convention designates flag state authority to monitor, investigate and penalize vessels that violate the obligations.

It is very important to monitor these oil spills to carry out emergency response clean-ups to reduce biological harm that can be caused to marine life. Under the Bonn Agreement,[349] monitoring procedures are set up to track oil spills and it requires evidence to prosecute ship owners. Satellite data in this instance plays a very prominent role in tracking oil spills to the vessel committing the offence.

- **Convention on Long-Range Transboundary Air Pollution (CLRTAP), 1979**

This aims to limit, reduce and prevent trans-boundary air pollution. Remote sensing can assist in collecting data concerning harmful gases and study the impact of air pollution on vegetation.

- **Montreal Protocol and Vienna Convention on Protection of the Ozone Layer, 1987**

This imposes a legal obligation to reduce or eliminate certain ozone-depleting substances. Satellite data under this protocol can assist in studying atmospheric ozone concentration and trace gases that harm the ozone layer.

---

spilled oil into Fildes Bay and Exxon Valdez spilling 11 million barrels of crude oil off the coast of Alaska. These incidents have led to killing of thousands of flora and fauna including the fishes.

[347] Raustiala, K., Reporting and Review Institutions in 10 Multilateral Environmental Agreements, Nairobi: UNEP, pp.19, 2001.

[348] Mitchell, R., Intentional Oil Pollution at Sea: Environmental Policy and Treaty Compliance, in: Choucri, N., Global Environmental Accord Series, Cambridge, MA: MIT Press, pp. 199, 1994.

[349] Agreement for Cooperation in Dealing with Pollution of the North Sea by Oil, Bonn, 9 June 1969, http://www.bonnagreement.org/about/history, (accessed 27.09.2019).

- **UN Convention on Biological Diversity (CBD), 1992**

This convention aims to conserve biological diversity and promote sustainable use. It obliges states to create national programs for the protection of biodiversity and to monitor threats. It requires states to set up protected areas to preserve biologically diverse habitats. Remote sensing can help study vegetation and wetlands. It helps in mapping habitat suitability to study the degraded ecosystem and monitor the recovery of threatened species.

- **UN Convention to Combat Desertification (CCD), 1992**

This aims to reduce desertification and droughts and provides standard self-reporting commitments by the parties. Remote sensing can help study desertification, drought patterns and vegetation covers. It also helps collect and analyse data to identify the causes.

- **Kyoto Protocol, 1997**

The Kyoto Protocol imposes legally binding commitments to limit or reduce the concentration of GHGs in the atmosphere. The main GHGs that the Protocol aims to reduce are carbon dioxide and methane. It allows for balancing of these emissions through an increase in vegetation; mainly through forests and wetlands. EO, as a result, plays a crucial role in providing information regarding forest areas, their type and density and monitoring of agricultural activities. Remote sensing helps support the implementation of the Kyoto Protocol. To monitor compliance of quantified data as provided under the Kyoto Protocol it specifies resolution standards by which forest areas are to be determined.[350]

As forest areas affect carbon storage, satellite data help determine and monitor forested areas. Satellite sensors can also monitor agricultural fields including rice fields that contribute heavily to global methane emissions. It helps study forest fires, which contribute significantly to global carbon stock. All this information is required to be included in the national inventories that are submitted under the Kyoto Protocol. EO is the only tool capable of providing information on land-use, land-cover and forest as required under the Protocol. Satellite image data helps in establishing the 1990 carbon stock baseline along with mapping and monitoring sources of methane.[351]

The Kyoto Protocol provides for monitoring compliance in the following areas that remote sensing can support:

- Provision of systematic observations of relevant land cover (Art. 5, 10),
- Support for the establishment of 1990 carbon stock baseline (Art. 3),
- Detection and spatial quantification of change in land cover (Art. 3, 12),
- Quantification of above-ground vegetation and

---

[350]Under Kyoto Protocol, forest areas must be determined using a spatial resolution no larger than 1 ha, corresponding to a satellite sensor resolution of less than 100 m.

[351]Niemeyer, I., et al., Treaty Monitoring from Space: Satellite Imagery Analysis for Verifying Treaty Compliance, Conference Paper in Proceedings of SPIE—The International Society for Optical Engineering, 2009.

- Mapping and monitoring of sources of methane.[352]

Remote sensing makes monitoring compliance an easier task as it allows gathering of satellite data consistently through systematic and repetitive observation over large areas. Remote sensing plays a vital role in monitoring compliance as it is an appropriate tool to gather land-cover information and estimate global terrestrial carbon storage required by the Protocol. Satellite data collected helps in studying the atmospheric trace gases, rise in sea level, and ice shelf melting, which assists in the further development of the Protocol.

To enable efficient and transparent monitoring, the Kyoto Protocol mandates 'National reporting' by parties as a compliance system. Annex I Parties to the protocol are to follow reporting procedures laid down to monitor compliance and are also required to report data annually. Considering the emission targets, the ability to furnish accurate data is a vital need. Accurate data of emissions is necessary as it helps determine what is available for trade. As Article 17 allows states to acquire emission units from other states, it is necessary to analyse information in the national reports as accurately as possible.[353] As data quality varies from state to state, use of remote sensing can help unify these differences and minimize poor data quality that casts doubt on the parties' act of compliance.[354]

Remote sensing is the most viable option available for states to monitor their compliance with the provisions of the Kyoto Protocol while simultaneously allowing them to collect national activity information. It requires all states to prepare a national inventory, which enables them to meet the reporting provisions under the treaty.[355] Remote sensing can also be utilized by independent bodies to verify the information submitted by states to ensure better compliance. It can also help collect data that is to be made available to the sensed state.

However, there are arguments that because treaty verification is not addressed in the Protocol, compelling states to fulfil obligations based on satellite data is not possible unless they themselves consent to monitoring by remote sensing. There is also a need for the required data to be centralised to ensure consistency.[356]

- **World Summit on Sustainable Development (WSSD), 2002**

---

[352] Ibid. *supra* note Peter, N., The Use of Remote Sensing to Support the Application of Multilateral Environmental Agreements, Space Policy, 20, pp. 193, 2004.

[353] UNFCCC, The Kyoto Protocol Mechanisms—International Emissions Trading Clean Development Mechanism Joint Implementation, https://unfccc.int/sites/default/files/cdm_kpm2010.pdf, (accessed 10.07.2019).

[354] Mullins, F., Kyoto Mechanism, Monitoring and Compliance—From Kyoto to The Hague, OECD Environment Directorate and International Energy Agency, COM/ENV/EPOC/IEA/SLT 9, pp. 37, 2001.

[355] Aschbacher, J., Monitoring Environmental Treaties Using Earth Observations, Verification Yearbook, Vertic, London, pp. 183, 2002 https://www.files.ethz.ch/isn/13529/VY02_FULL.pdf, (accessed 7.07.2019).

[356] Rosenqvist, A., A Review of Remote Sensing Technology in Support of the Kyoto Protocol, Environmental Science and Policy 6, pp. 451, 2003.

This requires states to ensure sustainability and work towards the eradication of poverty, access to clean water, sanitation, energy, health, trade and agriculture. To ensure that states meet these demands satellite data can be utilized to monitor land use and land cover, desertification, water resources and environmental degradation, and trace greenhouse gases (GHGs). The WSSD Plan of Implementation contains specific references to the utilization of remote sensing/EO as a crucial information source.[357]

- **UN Framework Convention on Climate Change (UNFCCC), 2002**

This requires states to take necessary actions to regulate GHGs and imposes legally binding targets to limit emissions. Remote sensing in this instance is used to measure land use, forestry, afforestation, reforestation and deforestation.

- **Paris Agreement, 2015**

The success of any international environmental agreement depends heavily on the ability and willingness of the parties to comply with their obligations under it. The Paris Agreement aims to reduce the increase in global temperatures to 2 degrees Celsius in this century. It also obligates states to pursue means to limit it to 1.5 degrees Celsius. It aims to create a transparent monitoring and reporting framework.[358] The main difference between the Kyoto Protocol and the Paris Agreement is the involvement of developing nations and their collective efforts to cut down on GHGs emission. Unlike the Kyoto Protocol, the Paris Agreement has no binding commitments and states can voluntarily decide on their emission targets. However, failing to meet the limits set by states for themselves does not attract penalties.

The important aspect of the Paris Agreement is monitoring, reporting and reassessing the commitments of states to move towards a better environment. The Agreement requires them to reassess and self-set targets for carbon reduction. 'Intended Nationally Determined Contributions' (INDC) allows states to set their own targets to be achieved.[359] Article 15 of the Paris Agreement establishes a mechanism of implementation and promotes compliance. Remote sensing could play a prominent role in the Paris Agreement but at national levels. As states volunteer to meet targets set for themselves, utilizing satellite data to furnish and support their commitment towards reaching the objectives of the agreement could play a prominent role.[360]

While there are obligations to reach a target, the Paris Agreement, however, mandates that every party is bound to submit a report on their commitment towards their

---

357 World Summit on Sustainable Development, Plan of Implementation, 4 September 2002, https://enb.iisd.org/2002/wssd/PlanFinal.pdf, (accessed 25.09.2019).

358 Denchak, M., Paris Climate Agreement: Everything You Need to Know, NRDC, 12 December 2018, https://www.nrdc.org/stories/paris-climate-agreement-everything-you-need-know, (accessed 10.07.2019).

359 Voigt, C., Operationalizing the Implementation and Compliance Mechanism in the Paris Agreement, Climate Law and Governance Initiative, 2016, http://www.climatelawgovernance.org/knowledgecenter/operationalizing-the-implementation-and-compliance-mechanism-in-the-paris-agreement/, (accessed 12.07.2019).

360 Voigt, C., The Compliance and Implementation Mechanism of the Paris Agreement, Review of European, Comparative & International Environmental Law, Vol. 25(2), pp. 161, 2016.

target. The Article 13 transparency framework gives the power to review each party's commitment and verify the reports. Article 14 aims to assess global commitment and progress towards achieving the Agreement's purpose.[361]

2.5.11.3 Issues Concerning the Use of Satellite Data to Monitor Compliance

Some argue that treaties are not automatically implemented in national legislation after their ratification. Implementation of treaties and their obligations is strongly dependent on national interests. Because treaties are not signed by non-state entities, implementing and monitoring compliance becomes very limited.[362]

While there are requirements for different types of remote sensing technology and satellite imagery based on the obligations of the treaties to be fulfilled, their contributions can be lacking in certain aspects as there is no consistency and common interest between some of the treaties.[363] Remote sensing by itself is not sufficient to provide all the required data for monitoring compliance, hence making compliance dependent on other sources for data collection. The absence of compliance monitoring mechanisms in most treaties results in the limited information available to evaluate monitoring aspects. Also, remote sensing is sometimes not regarded as a substantive source of evidence as outcomes depend on the type of instruments used for collecting the data. This calls for transparency in the tools used, methodologies, sources and capacities.

Satellite data also faces the issue of the legal validity of the data collected. Though UN Principles on Remote Sensing allow and encourage the use of such data, states have been questioning the legitimacy of such data collected. Satellite data are usually criticized for lack of consistency and standardization issues along with fragmented and inadequate data archives. The use of satellite data requires a competent authority to verify it as per the treaty compliance requirements. There is also a need for treaties to specifically provide for the use of satellite data as a tool to monitor compliance.[364]

---

[361] Biniaz, S., Elaborating Article 15 of the Paris Agreement: Facilitating Implementation and Promoting Compliance, Policy Brief IDDRI, pp. 2, 2017, https://www.iddri.org/sites/default/files/import/publications/pb1017_art-15-pa_sue-biniaz.pdf, (accessed 12.07.2019).

[362] Vollmer, R., Summary and Recommendations, in: Wirkus, L., Vollmer, R., Monitoring Environment and Security—Integrating Concepts and Enhancing Methodologies, BICC Seminar Documentation, Brief 37, pp. 74, https://www.researchgate.net/publication/283321139_Monitoring_Environment_and_Security_-_Integrating_concepts_and_enhancing_methodologies, (accessed 8 July 2019).

[363] Ibid. *supra* note Purdy, R., Using Earth Observation Technologies for Better Regulatory Compliance and Enforcement of Environmental Law, Journal of Environmental law 22:1, pp. 64, 2010.

[364] Irmgard Niemeyer, "*Challenges in Treaty Monitoring*," In: Lars Wirkus and Ruth Vollmer, "*Monitoring Environment and Security—Integrating Concepts and Enhancing Methodologies*," BICC Seminar Documentation, Brief 37 at 66 https://www.researchgate.net/publication/283321139_Monitoring_Environment_and_Security_-_Integrating_concepts_and_enhancing_methodologies, (accessed 8 July 2019).

Chapter 3
Use of Satellite Data in Courts

Abstract This chapter will provide a detailed analysis of satellite data used as evidence in international courts. A short description of the International Court of Justice and the International Criminal Court will be provided, and individual cases will be analysed. The aim is to underline and prove how satellite data is useful as evidence in international courts. The analysed cases will include territorial and maritime disputes, maritime delimitation, prosecution of serious international crimes such as genocide, war crimes and crimes against humanity. Also, this chapter contains the detailed explanations of a satellite imagery expert in interpreting the satellite images provided by the parties and the need for judges to understand the procedure to produce satellite images that are used as evidence. Further, some aspects of small satellite programs and the current use of space telecommunication infrastructure in given countries will be analysed. The focus will be on highlighting the growing demand for space activities in Africa, particularizing the activities of the most active countries, but also the creation of the African Space Agency.

3.1 Space Endorsement for Courts and Developing Countries

The International Court of Justice (ICJ) is different from the International Criminal Court (ICC), principally because the ICJ settles international legal disputes between states, while the ICC tries individuals charged with genocide, war crimes, crimes against humanity and the crime of aggression. However, satellite imagery has proved useful in various cases in both jurisdictions.

3.1.1 The International Court of Justice

The International Court of Justice (ICJ) is the principal judicial organ of the UN.[1] According to the Charter of the UN, each member of the UN must comply with the

[1]Article 92 of the Charter of the United Nations, https://treaties.un.org/doc/publication/ctc/uncharter.pdf, (accessed 07.07.2019).

A. Froehlich and C. M. Tăiatu, *Space in Support of Human Rights*,
Studies in Space Policy 23, https://doi.org/10.1007/978-3-030-35426-8_3

decisions of the ICJ in any case in which it is a party. Should a party fail to comply with a ruling, the other party may have recourse to the Security Council.[2] The Statute of the ICJ is an integral part of the UN Charter and provides that only states may be parties in cases before this court. In addition, the Statute provides that the ICJ may request relevant information from public international organizations and may even receive such information presented by such organizations on their own initiative.[3]

The jurisdiction of the ICJ comprises all matters specially provided for in the UN Charter or in treaties and conventions in force. According to Article 38 of the ICJ Statute, the Court, whose function is "to decide in accordance with international law such disputes as are submitted to it" shall apply:

> a) International conventions, whether general or particular, establishing rules expressly recognized by the contesting states; b) International custom, as evidence of a general practice accepted as law; c) the general principles of law recognized by civilized nations; d) subject to the provisions of Article 59, judicial decisions and the teachings of the most highly qualified publicists of the various nations, as subsidiary means for the determination of rules of law.

The ICJ has contributed to the development of human rights by providing solutions to escalating international problems, condemning violations of rights such as terrorism and the unilateral use of force.[4]

For the purpose of proving the usefulness of satellite imagery and underlining the questions that are being raised in practice with respect to probatory value, the practice of the ICJ will be analysed further, highlighting the cases below.

3.1.1.1 International Court of Justice: Territorial and Maritime Dispute Between Nicaragua and Honduras in the Caribbean Sea

On 8 December 1999, Nicaragua filled an application against Honduras, requesting the ICJ to decide on the maritime delimitation appertaining to the states in the Caribbean Sea. The ICJ proceeded to determine the delimitation itself from the analysis of the materials provided by both states, and found that no established boundary existed along the 15th parallel on the basis of either *uti possidetis juris* or tacit agreement between the parties.[5] The judgement of 8 October 2007 revealed that both states had provided satellite images to the ICJ.

The Agent of Nicaragua informed the ICJ that its government understood how to make use of satellite images to support its arguments. Because the satellite images

---

[2] Article 94 of the Charter of the United Nations.

[3] Article 34, Chap. 2 "Competence of the Court" of the Statute of the International Court of Justice, https://treaties.un.org/doc/publication/ctc/uncharter.pdf, (accessed 07.07.2019).

[4] Bedi, S.R.S., The Development of Human Rights Law by the Judges of the International Court of Justice, Oxford, Portland Oregon. Hart Publishing, 2007.

[5] *Uti possidetis juris* is the principle explaining that the internal administrative boundaries of the pre-independence sovereign, in this case Spain, form the international boundaries of the several successor states, here including Nicaragua and Honduras, https://scholarship.law.duke.edu/cgi/viewcontent.cgi?article=2553&context=faculty_scholarship, (accessed 24.09.2019).

were "part of a publication readily available" pursuant to paragraph 4 of Article 56 of the Rules of the ICJ, they could be referred to during the oral proceedings.[6] Honduras used satellite images to support its oral argument about the mainland Honduran and Nicaraguan coasts at Cape Gracias a Dios. The Memorial of the Government of Nicaragua used satellite images to prove "the river mouth shows further movement seaward towards the north and east than was the case in 1962".[7] The Government of Nicaragua used satellite imagery to prove the mouth of the Coco River travelled more than 1 nautical mile in a north easterly direction.[8] A satellite picture was included in Figure VII of the memorial. Separately, in its Memorial Honduras "shared the view expressed in the Nicaraguan Memorial" when referring to the geographical changes, in particular about the mouth of the rivers that had moved eastwards.[9] The evidence analysed by the Honduras Government was a series of LANDSAT satellite images.

The ICJ however did not authorize the Government of Nicaragua to produce the remaining documents for procedural reasons.[10] Also, Honduras stated that its Satellite Operation Plan was related to drug enforcement operations, and involved reconnaissance operations to identify possible targets, areas and installations used or connected to drug trafficking.[11] The relevance of this plan related to the fact that the list of "islets and cays" authorized for the operation included Boble Cay, South Cay, Half Moon Cay and Savanna Cay, which were disputed at the ICJ. Nicaragua objected. Another example of such evidence was the satellite photograph used by Honduras to present the provisional equidistance line to prove the endpoint of the mainland Honduran and Nicaraguan coasts at Cape Gracias a Dios.[12]

The ICJ found unanimously that the Republic of Honduras had sovereignty over Bobel Cay, Savanna Cay, Port Royal Cay and South Cay and decided the starting-point of the single maritime boundary that divides the territorial sea, continental shelf and Exclusive Economic Zones (EEZ) of the Republic of Nicaragua and the Republic of Honduras.[13] To draw a general conclusion, the particularity of decisions in relation to maritime delimitation is that they depend largely on the geomorphological circumstances of the case. In this case concerning a territorial and maritime dispute between Nicaragua and Honduras, the Court took into consideration the changing

[6] Judgment of 8 October 2007 "Territorial and Maritime Dispute between Nicaragua and Honduras in the Caribbean Sea (Nicaragua v. Honduras), https://www.icj-cij.org/en/case/120, (accessed 9.06.2019).

[7] Memorial submitted by the Government of Nicaragua, Volume I, Paragraph 25, pp. 12, 2001, https://www.icj-cij.org/files/case-related/120/13719.pdf, (accessed 10.06.2019).

[8] Ibid. *supra* note Memorial submitted by the Government of Nicaragua, Volume I, Paragraph 23, pp. 158, 2001.

[9] Counter-Memorial of the Republic of Honduras, Chap. 2, (D) The instability of the Mouth of the River Coco, Paragraph 7.12. pp. 136.

[10] Judgment 8 October 2007, Paragraph 12, pp. 10, https://www.icj-cij.org/files/case-related/120/120-20071008-JUD-01-00-EN.pdf, (accessed 24.09.2019).

[11] Ibid. *supra* note Judgment 8 October 2007, Paragraph 183, pp. 59.

[12] Ibid. *supra* note Judgment 8 October 2007, Paragraph 276, pp. 87.

[13] Ibid. *supra* note Judgment 8 October 2007, Paragraph 321, pp. 105.

shape of the River Coco delta which determined the method to be used in drawing a delimitation line. It was also relevant to acknowledge that the separation on each side of the margin of the River Coco was determined by the roughly perpendicular and convex coasts.[14]

3.1.1.2 International Court of Justice: Maritime Delimitation in the Caribbean Sea and the Pacific Ocean (Costa Rica v. Nicaragua)

On 25 February 2014, the Republic of Costa Rica (Costa Rica) filed an Application with the International Court of Justice (ICJ) against the Republic of Nicaragua (Nicaragua) requesting the court: *"to determine the complete course of a single maritime boundary between all the maritime areas appertaining, respectively, to Costa Rica and to Nicaragua in the Caribbean Sea and in the Pacific Ocean, on the basis on international law"*. Costa Rica requested the ICJ *"to determine the precise geographical co-ordinates of the single maritime boundaries in the Caribbean Sea and in the Pacific Ocean"* and claimed that the starting-point of the maritime delimitation between the Parties on the Caribbean side was "on the right bank of the San Juan River at its mouth". Nicaragua contended that the starting-point was situated at the extremity of Punta de Castilla, near the north-eastern corner of Harbor Head Lagoon, 3.59 km east of that suggested by Costa Rica. The Court upheld Nicaragua's claim concerning sovereignty over the northern coast of Isla Portillos. The Court further decided that Costa Rica had sovereignty over the whole northern part of Isla Portillos, with the exception of Harbor Head Lagoon and the sandbar separating it from the Caribbean Sea, whose sovereignty pertains to Nicaragua. The ruling included a decision on the maritime boundary and the obligation of Nicaragua to remove a military camp.[15]

This case was challenging because according to Nicaragua and Costa Rica there was "a series of discontinuous coast-parallel lagoons in a coastal stretch of Isla Portillos, between the beach and an area covered tree vegetation."[16] Nicaragua highlighted some characteristics of the river with the help of the satellite image provided

---

[14] Barker, C.J., Decisions of International Courts and Tribunals, https://www.jstor.org/stable/20488238?seq=1#page_scan_tab_contents, (accessed 24.09.2019).

[15] International Court of Justice, Maritime Delimitation in the Caribbean Sea, and the Pacific Ocean (Costa Rica v. Nicaragua) and Land Boundary in the Northern Part of Isla Portillos (Costa Rica v. Nicaragua), paragraph 205, 2 February 2018, https://www.icj-cij.org/files/case-related/157/157-20180202-JUD-01-00-EN.pdf, (accessed 24.09.2019).

[16] Comment 4, Response to the Comments of Costa Rica on the Report submitted on 30 April 2017 by the experts appointed by Court in the Case concerning Maritime Delimitation in the Caribbean Sea and the Pacific Ocean (Costa Rica v. Nicaragua), Paragraph 106, Figs. 41–42, pp. 2, https://www.icj-cij.org/files/case-related/157/157-20170608-WRI-01-00-EN.pdf, (accessed 11.06.2019).

by LANDSAT in 2011 showing a large abandoned channel of Taura River connected with the San Juan River that used to flow into the Caribbean Sea.[17]

The experts extensively used satellite images when analysing the site for geographical and geomorphological setting.[18] The Report of Experts from 30 April 2017, combined information from the field acquired from two site visits.[19] The first visit was conducted in December 2016 after Hurricane Otto impacted the area. The second visit was in March when there was the lowest average rainfall in the area.[20] The findings in the field were presented to the Court with the help of annotated satellite images provided by the U.S. Geological Survey. They showed the location of San Juan del Norte village, Rio Indio Lodge, and that the position of the San Juan River and its mouth was highly variable.[21] Drones provided by Nicaragua were used to take pictures and videos of the coast.[22] Another satellite image provided by Digital Globe was used to analyse from above the sky the area of the airport with locations of the corners of the cemeteries and church steps that were previously investigated by foot.[23] Aerial photograph was also used to determine the area of the river mouth, when the sand spit was largely diminished.[24] Another satellite image from January 2016 was used to analyse the area of the lagoon with the points of interest measured during the two site visits. The points of interest were projected on a satellite image with a number of co-ordinates along the boundaries of the search area, containing (i) points measured during the first site visit; (ii) points measured during

---

[17] The lower reach of the San Juan River has favorable geomorphological conditions for experiencing significant changes in its path. The channel is not subject to stabilization or control measures and flows along a flat and unconfined coastal plain locally carved by abandoned channels that may reactivate during flood events. This concept is supported by the presence of a large abandoned channel (Taura River) indicating that the configuration of the river course has been affected by major changes in historical times (see Figs. 1 and 2 below). Moreover, if coastal recession proceeds in the future at a similar pace as in the recent past, then the eastward shift in the mouth of the San Juan River, related to the intersection between the retreating coastline and the meander located around 300 m inland, is rather likely, in the mid-term, especially considering the current sea-level rise scenario (see Fig. 88 of the Report). For instance, the coast has retreated around 940 m on the east side of Los Portillos/Harbor Head Lagoon between 1898 and 2009, yielding an average retreat rate of around 8.5 m/yr (see paragraph 192 and Fig. 86 of the Report).

[18] Foauche, E., Gutierrez, F., Experts opinion, Maritime Delimitation in the Caribbean Sea (Costa Rica v. Nicaragua), 2017, https://www.icj-cij.org/files/case-related/157/157-20170430-WRI-01-00-EN.pdf, (accessed 11.06.2019).

[19] Ibid. *supra* note Foauche, E., Gutierrez, F., Experts opinion, Maritime Delimitation in the Caribbean Sea (Costa Rica v. Nicaragua), 2017.

[20] Ibid. *supra* note Foauche, E., Gutierrez, F., Experts opinion, Maritime Delimitation in the Caribbean Sea (Costa Rica v. Nicaragua), Paragraph 8–10, 2017.

[21] Ibid. *supra* note Foauche, E., Gutierrez, F., Experts opinion, Maritime Delimitation in the Caribbean Sea (Costa Rica v. Nicaragua), Fig. 1, 2017.

[22] Ibid. *supra* note Ibid. *supra* note Foauche, E., Gutierrez, F., Experts opinion, Maritime Delimitation in the Caribbean Sea (Costa Rica v. Nicaragua), Paragraph 18.

[23] Ibid. *supra* note Foauche, E., Gutierrez, F., Experts opinion, Maritime Delimitation in the Caribbean Sea (Costa Rica v. Nicaragua), 2017, Fig. 9.

[24] Ibid. *supra* note Foauche, E., Gutierrez, F., Experts opinion, Maritime Delimitation in the Caribbean Sea (Costa Rica v. Nicaragua), 2017, Fig. 13.

the second site visit; (iii) the area investigated with metal detectors and iron rods and (iv) the area investigated with Ground-Penetrating Radar (GPR).[25] The Report also acknowledged that experts had made use of aerial photographs, maps, satellite images, geodetic data, tide records and discharge time series before conducting the first site visit.[26]

Also, for the geographical and geomorphological setting, detailing the starting-points of the maritime boundary between Nicaragua and Costa Rica in the Caribbean Coast, the experts used an annotated satellite image provided by LANDSAT in December 2016. The image contained the main geographical-geomorphic features of the lower reaches of the San Juan fluvial system associated with the Caribbean coast, showing that around 19 km west of the Caribbean coast near Delta, the San Juan River splits into two main channels.[27] The image combined with site visits helped to establish the effect of erosional coastal processes. Comparing the evolution of the coastline, the experts used a cartographic overlay of the coastline as represented in an historical map of 1840, on a satellite image taken on October 2013.[28] A drone of the Costa Rican delegation provided an image of Los Portillos/Harbour Head Lagoon on March 2017 during the second visit of experts.[29] Also the Nicaraguan Delegation provided evidence with drones.[30]

As highlighted in the conclusions of the Report, the experts used satellite images to advise the ICJ. The mission of the experts was to "advise the Court regarding the state of the coast between the point suggested by Costa Rica and the point suggested by Nicaragua in their pleadings as the starting-point of the maritime boundary in the Caribbean Sea" and to answer four specific questions when analysing the site for geographical and geomorphological setting.[31]

(a) What are the geographical co-ordinates of the point at which the right bank of the San Juan River meets the sea at the low-water line?
(b) What are the geographical co-ordinates of the land point which most closely approximates to that identified by the first Alexander Award as the starting-point of the land boundary?
(c) Is there a bank of sand or any maritime feature between the points referred to in subparagraphs (a) and (b) above? If so, what are their physical characteristics?

[25] Ibid. *supra* note Foauche, E., Gutierrez, F., Experts opinion, Maritime Delimitation in the Caribbean Sea (Costa Rica v. Nicaragua), Fig. 18, 2017.

[26] Ibid. *supra* note Foauche, E., Gutierrez, F., Experts opinion, Maritime Delimitation in the Caribbean Sea (Costa Rica v. Nicaragua), Paragraph 85, 2017.

[27] Ibid. *supra* note Foauche, E., Gutierrez, F., Experts opinion, Maritime Delimitation in the Caribbean Sea (Costa Rica v. Nicaragua), Fig. 19, 2017.

[28] Ibid. *supra* note Foauche, E., Gutierrez, F., Experts opinion, Maritime Delimitation in the Caribbean Sea (Costa Rica v. Nicaragua), Fig. 22, 2017.

[29] Ibid. *supra* note Foauche, E., Gutierrez, F., Experts opinion, Maritime Delimitation in the Caribbean Sea (Costa Rica v. Nicaragua), Figs. 33, 36, 42–43 and 47, 2017.

[30] Ibid. *supra* note Foauche, E., Gutierrez, F., Experts opinion, Maritime Delimitation in the Caribbean Sea (Costa Rica v. Nicaragua), Fig. 46, 2017.

[31] Ibid. *supra* note Foauche, E., Gutierrez, F., Experts opinion, Maritime Delimitation in the Caribbean Sea (Costa Rica v. Nicaragua), Paragraph 111, 2017.

In particular, are these features, or some of them, permanently above water, even at high tide? Is Los Portillos/Harbor Head Lagoon separated from the sea?

(d) To what extent is it possible, or probable, that the area concerned will undergo major physical changes in the short and long term?

Regarding the first and second question, Costa Rica submitted a Memorial claiming that the starting point of the maritime-delimitation between the parties on the Caribbean side was ""*on the right bank of the San Juan River at its mouth: the point at which the line dividing the land territories of the two States intersects the coast. That point is located at the north-western extremity of Costa Rica's Isla Portillos, where Costa Rica's land territory and Nicaragua's waters of the San Juan River meet the Caribbean Sea" (Memorial of Costa Rica, para. 4.13)*" and places the starting-point of the delimitation ""*at the base of the sand spit extending northwest from Isla Portillos, because no reliable basepoints can be derived from this ephemeral low-lying feature. The co-ordinates of the starting point in the Caribbean ... are 10° 56′ 26.0″ N and 83° 41′ 53.0″ W" (Memorial of Costa Rica, para. 4.15)."*

Nicaragua submitted a Counter-Memorial arguing that this point was to be found about 3.6 km east of that suggested by Costa Rica (ibid. paras. 3.35, 3.48, 3.52). Recalling the 1858 Treaty of Limits, the 1888 Cleveland Award and the Alexander Awards, it considered that the starting-point of the land boundary (and the maritime delimitation) was the extremity of Punta de Castilla, which it placed near the north-eastern corner of Los Portillos/Harbor Head Lagoon. According to Nicaragua, the starting-point of the sea delimitation was located at 10° 55′ 49.7″ North and 83° 40′ 0.6″ West (ibid. paras. 3.38–3.48, 3.52).

The Report of experts focused on the points defined above by the delegations of Costa Rica and Nicaragua. The experts used the annotated satellite image from June 2013 provided by DIGITAL GLOBE and the U.S. Geological Survey to indicate to the ICJ the approximate location of the starting points of the maritime boundary proposed by Costa Rica and Nicaragua.[32]

Another satellite image from 2003 was used by the experts to project the coordinates measured by Nicaragua and Costa Rica for the markers found in 2003 "in the beach (square representing average coordinates and triangles the measurements of the Parties) and submerged in the lagoon (yellow circle)".[33]

The satellite image of 18 December 2009, which was "acquired by the RAPIDEYE sensor, and already georeferenced when purchased from DIGITAL GLOBE" was used by the experts as a basemap in the georeferencing procedures. In particular, the satellite image was used as basemap for georeferencing the Map of Greytown

[32] Ibid. *supra* note Foauche, E., Gutierrez, F., Experts opinion, Maritime Delimitation in the Caribbean Sea (Costa Rica v. Nicaragua), Fig. 48, 2017.

[33] Ibid. *supra* note Foauche, E., Gutierrez, F., Experts opinion, Maritime Delimitation in the Caribbean Sea (Costa Rica v. Nicaragua), Fig. 60, 2017.

Table 1 Table presented by the satellite experts to the ICJ, containing satellite images provided by Digital Globe

Acquisition date	Satellite sensor	Spatial resolution (m)
18 December 2009	RAPIDEYE	5
10 March 2011	WORLDVIEW2	0.5
26 November 2013	WORLDVIEW 2	0.5
4 March 2014	WORLDVIEW 2	0.5
26 September 2014	WORLDVIEW 2	0.5
22 January 2016	WORLDVIEW 3	0.5

Harbor of 1899[34] and the Ground Control Points used for georeferencing.[35] Another example of the contribution of satellite images as evidence in courts is the transparent overlay of the Map of Greytown Harbor of 1899 applied on the satellite image of 2009. Such procedure was used to show the alignment of the corners of the cemeteries measured with GPS by the parties, depicting the estimated centre of Plaza Victoria and the linear edge of the high vegetation south of the landing strip.[36] Overlays of the georeferenced Map of Greytown of 1888 on the 1961 aerial image (provided by Costa Rica on 20 October 2016) and the 2009 satellite image were used by the experts to provide information to the ICJ.[37] Various other measurements were made with the help of the transparent overlay of the sketch on the satellite image of 2009 including the centre of Plaza Victoria and the triangulation data included in the Proceedings of the Costa Rica-Nicaragua Demarcation Commissions.[38] Recently available satellite data, such as the satellite image of 22 January 2016, was used to measure the land point that most closely approximates the Initial Marker.[39]

Regarding the third question, the experts used satellite images provided by DIGITAL GLOBE from multiples dates to analyse and identify geomorphic features along the coastline, such as mentioned in the following table presented by experts in the Report for the ICJ (Table 1).[40]

[34]Ibid. *supra* note Foauche, E., Gutierrez, F., Experts opinion, Maritime Delimitation in the Caribbean Sea (Costa Rica v. Nicaragua), Fig. 68, 2017.

[35]Ibid. *supra* note Foauche, E., Gutierrez, F., Experts opinion, Maritime Delimitation in the Caribbean Sea (Costa Rica v. Nicaragua), Fig. 69, 2017.

[36]Ibid. *supra* note Foauche, E., Gutierrez, F., Experts opinion, Maritime Delimitation in the Caribbean Sea (Costa Rica v. Nicaragua), Figs. 72–73, 2017.

[37]Ibid. *supra* note Foauche, E., Gutierrez, F., Experts opinion, Maritime Delimitation in the Caribbean Sea (Costa Rica v. Nicaragua), Figs. 76, 77, 78 and 79, 2017.

[38]Ibid. *supra* note Foauche, E., Gutierrez, F., Experts opinion, Maritime Delimitation in the Caribbean Sea (Costa Rica v. Nicaragua), Figs. 81, 82, 2017.

[39]Ibid. *supra* note Foauche, E., Gutierrez, F., Experts opinion, Maritime Delimitation in the Caribbean Sea (Costa Rica v. Nicaragua), Figs. 83, 84 and 85, 2017.

[40]Ibid. *supra* note Foauche, E., Gutierrez, F., Experts opinion, Maritime Delimitation in the Caribbean Sea (Costa Rica v. Nicaragua), Table 3 "Satellite images provided by Digital Globe", pp. 73, 2017.

The experts concluded that the observations carried out during the site visits and the available data from satellite images and aerial photos, indicated that Los Portillos/Harbour Head Lagoon is commonly separated from the sea by the sand barrier. Only in the case of high energy waves related to storms or hurricanes can there be a temporary breach of the natural barrier allowing water to circulate between the lagoon and the sea.[41] In addition, satellite images also revealed the presence of coast-parallel shoals.[42]

In response to the fourth question, the experts georeferenced the cartographic sketch included in Proceedings X of the Costa Rica-Nicaragua Demarcation Commissions from 1898, using different coordinates and markers. As mentioned in the Report, "The comparison of the georeferenced sketch with a satellite image from 2009 allows to roughly estimate a coastal retreat of 940 m on the east side of Los Portillos/Harbour Head Lagoon, yielding an average recession rate of around 8.5 m/yr".[43] The usefulness of the satellite data in comparing historical maps with recent observations was consistently demonstrated. Costa Rica's comments on the Expert Report stated that the course of the lower reach of San Juan river, up to a point just before its mouth, had been relatively stable from a geomorphological point of view.[44] The experts response to the comments of Costa Rica highlighted that "the lower reach of the San Juan River has favourable geomorphological conditions for experiencing significant changes in its path" mentioning the lack of any stabilization or control measures and the abandoned channels that may reactivate during flood events.[45]

3.1.1.3 Judgment on the Status of the Kasikili/Sedudu Islands Solving a Territorial Dispute

The case concerned the boundary between the Republic of Botswana ("Botswana") and the Republic of Namibia ("Namibia") around Kasikili/Sedudu Island located in Chobe River (In Namibia is known as Kasikili and in Botswana as Sedudu). On 24 May 1992, Botswana and Namibia appointed a Joint Team of Technical Experts do determine the boundary between these two countries, around Kasikili/Sedudu Island. The joint team had to follow the Treaty of 1 July 1890 signed between Great Britain

---

[41] Ibid. *supra* note Foauche, E., Gutierrez, F., Experts opinion, Maritime Delimitation in the Caribbean Sea (Costa Rica v. Nicaragua), Paragraph 187, pp. 73, 2017.

[42] Ibid. *supra* note Foauche, E., Gutierrez, F., Experts opinion, Maritime Delimitation in the Caribbean Sea (Costa Rica v. Nicaragua), Paragraph 190, pp. 74, 2017.

[43] Ibid. *supra* note Foauche, E., Gutierrez, F., Experts opinion, Maritime Delimitation in the Caribbean Sea (Costa Rica v. Nicaragua), Fig. 86. Paragraph 192, pp. 74, 2017, https://www.icj-cij.org/files/case-related/157/157-20170430-WRI-01-00-EN.pdf, (accessed 24.09.2019).

[44] Observations of Costa Rica upon the report prepared by the Court-appointed experts, pp. 5, https://www.icj-cij.org/files/case-related/157/157-20170601-WRI-01-00-EN.pdf, (accessed 14.06.2019).

[45] Response to the Comments of Costa Rica on the Report submitted on 30 April 2017 by the experts appointed by the Court in the Case concerning Maritime Delimitation in the Caribbean Sea and the Pacific Ocean (Costa Rica v. Nicaragua), Comment 6, pp. 3, https://www.icj-cij.org/files/case-related/157/157-20170608-WRI-01-00-EN.pdf, (accessed 14.06.2019).

and Germany regarding the spheres of influence of the two countries in Africa and the applicable principles of international law. The Joint Team of Technical Experts was unable to reach a common conclusion. Therefore, on 17 May 1996, Botswana and Namibia transmitted a joint letter to the ICJ in order to settle the dispute. The ICJ was asked to determine the boundary and the legal status of the island on the basis of the Anglo-German treaty of 1 July 1890 and the principles of international law. In the Courts opinion, "the real dispute between the Parties concerns the location of that main channel where the boundary lies".[46] Botswana is contending that it was the channel running north of Kasikili/Sedudu Island and Namibia that is the channel running south of the island. Thus, the Court proceeded in determining which was the main channel of the Chobe river around the island in question.

Both parties to the dispute used the satellite imagery in support of their judicial position. Botswana provided satellite imagery to prove the main channel was the northern channel because is wider and deeper than the southern channel. A comparison was made using satellite images from June 1975, March 1995 and June 1996, both during the dry and rainy seasons.[47] Satellite imagery was not directly mentioned in the final judgement but satellite data was reasonably useful to complement the evidence and support the court judgement which decided that "the boundary (…) follows the line of deepest soundings in the northern channel of the Chobe River around Kasikili/Sedudu Island" and also that "Kasikili/Sedudu Island forms part of the territory of the Republic of Botswana".[48] Because there were no available satellite images from the beginning of the agreement in question, the Court used images that prove relevant in relation to the question whether Kasikili Island was occupied or cultivated.[49]

The ICJ decided with eleven votes to four regarding the boundary between Botswana and Namibia around Kasikily/Sedudu Island, considering the argument providing that Namibia had prescriptive title to Kasikili/Sedudu Island resulted from the exercise of sovereign jurisdiction with full knowledge and acceptance by the authorities of Botswana and its predecessors. The judgement of 13 December 1999 contains four dissenting opinions of the judges. For example, in the opinion of Judge Parra-Aranguren, the aerial photographs and satellite images submitted to the ICJ as evidence, are not relevant for the case because they were taken after 1914 and the Treaty invoked by both Parties was concluded in 1890. Relevant for this case is that the separate opinions of the judges did not result from the technical interpretation of

[46]ICJ -International Court of Justice, Judgment of 13 December 1999, Paragraph 25, pp. 21 https://www.icj-cij.org/files/case-related/98/098-19991213-JUD-01-00-EN.pdf, (accessed 06.09.2019).

[47]Ibid. *supra* note ICJ -International Court of Justice, Judgment of 13 December 1999, Paragraph 25, pp. 21 https://www.icj-cij.org/files/case-related/98/098-19991213-JUD-01-00-EN.pdf, (accessed 06.09.2019).

[48]Ibid. *supra* note ICJ -International Court of Justice, Judgment of 13 December 1999, Paragraph 25, pp. 21 https://www.icj-cij.org/files/case-related/98/098-19991213-JUD-01-00-EN.pdf, (accessed 06.09.2019).

[49]Ibid. *supra* note ICJ -International Court of Justice, Judgment of 13 December 1999, Paragraph 25, pp. 21 https://www.icj-cij.org/files/case-related/98/098-19991213-JUD-01-00-EN.pdf, (accessed 06.09.2019).

satellite imagery used as evidence.[50] The ICJ decided unanimously that the nationals of, and vessels flying the flags of, Botswana and Namibia, shall enjoy equal national treatment.[51]

3.1.2 *The International Criminal Court*

The International Criminal Court (ICC) was established according to the Rome Statute (Statute) of the International Criminal Court originally circulated as document A/CONF.183/9 of 17 July 1998 and corrected by *procès-verbaux* of 10 November 1998, 12 July 1999, 30 November 1999, 8 May 2000, 17 January 2001 and 16 January 2002.[52] The seat of the ICC was established in The Hague, making the Netherlands the host country. The ICC is the only international court of justice that has the authority to prosecute "the most serious crimes of international concern" and it is complementary to national criminal jurisdictions.[53] The relationship of the ICC to the UN created specific dynamics and operational links regarding international relations, being approved by official cooperative arrangements which are especially useful in pursuing investigations on areas of tremendous turmoil, conflict and death.[54]

According to Article 5 of the ICC Statute referring to jurisdiction, admissibility and applicable law, the ICC limits its jurisdiction to international crimes, the "most serious crimes of concern to the international community as a whole" such as:

(a) The crime of genocide; (Article 6)
(b) Crimes against humanity; (Article 7)
(c) War crimes; (Article 8)
(d) The crime of aggression.[55] (Article 8 bis)

It has been underlined that "the Statute breaks new ground in defining international criminal acts" which refers to crimes against humanity which, even if existing under this concept, no international treaty had defined until the ICC statute did so.[56] Article 8 of the ICC Statute includes three categories of war crimes, referring to: (i) Grave breaches of the Geneva Conventions of 1949 with an emphasis on several violations;

---

[50] WorldCourts, Judgement Kasikili/Sedud Island, Paragraph 78, 13 December 1999, http://www.worldcourts.com/icj/eng/decisions/1999.12.13_kasikili.htm, (accessed 25.09.2019).

[51] Ibid. *supra* note ICJ -International Court of Justice, Judgment of 13 December 1999, Paragraph 25, pp. 67 https://www.icj-cij.org/files/case-related/98/098-19991213-JUD-01-00-EN.pdf, (accessed 06.09.2019).

[52] Done at Rome on 17 July 1998, in force on 1 July 2002, United Nations, Treaty Series, Vol. 2187, No. 38544, Depositary: Secretary-General of the United Nations, http://treaties.un.org, (accessed 24.09.2019).

[53] Ibid. *supra* note. Rome 1998 Statute. *Article* 1 "The Court".

[54] Schiff, B.N., Building the International Criminal Court, Cambridge University Press, pp. 1, 2008.

[55] Ibid. *supra* note. Rome 1998 Statute. Article 5 "Crimes within the jurisdiction of the Court".

[56] Schiff, B.N., Building the International Criminal Court, Cambridge University Press, pp. 75, 2008.

(ii) Other serious violations of the laws and customs applicable in international armed conflict, within the established framework of international law with a particular emphasis on a number of violations; and (iii) the crimes that pertain "in case of an armed conflict *not of an international character*" with an emphasis on a number of violations.[57]

When referring to the duties and powers of the Prosecutor with respect to investigations, the Statute provides that the Prosecutor, in establishing the truth, may extend the investigation to cover all facts and evidence relevant to an assessment of whether there is criminal responsibility and should take any appropriate measures to ensure the effective investigation and prosecution of crimes within the jurisdiction of the ICC.[58] The role of the Prosecutor includes the collection and examination of evidence, for which the Prosecutor may seek cooperation with IGOs or other arrangement that would support gathering of evidence.[59]

The ICC Statute contains rules about the appointment of experts. Article 56, when referring to the Role of the Pre-Trial Chamber in relation to a unique investigative opportunity, provides that, upon the request of the Prosecutor and the necessity to ensure the efficiency and integrity of the proceedings, measures may include appointing an expert to assist. Also, Article 93, referring to other form of cooperation, provides that state parties shall, in accordance with the provisions of the Statute and national law, provide assistance to the ICC by producing evidence, including expert opinions and reports necessary to the investigation, or facilitating the appearance of the experts before the ICC. Art 100 provides that the state shall support the ordinary costs for execution of a request in its territory except that the ICC shall bear the costs of any expert opinion or report requested by the ICC.[60] The ICC Statute refers to the Rules of Procedure and Evidence (Rules) and provides the procedure for their adoption and amendment. According to Article 51 of the ICC Statute, the Rules are subordinate to the ICC Statute and their role is to provide clarification

---

[57] Ibid. *supra* note. Rome 1998 Statute, Article 8, https://www.icc-cpi.int/resource-library/Documents/RS-Eng.pdf, (accessed 25.09.2019).

[58] Ibid. *supra* note. Rome 1998 Statute. *Article 54.*

[59] (a) Collect and examine evidence;
(b) Request the presence of and question persons being investigated, victims and witnesses;
(c) Seek the cooperation of any State or intergovernmental organization or arrangement in accordance with its respective competence and/or mandate;
(d) Enter into such arrangements or agreements, not inconsistent with this Statute, as may be necessary to facilitate the cooperation of a State, intergovernmental organization or person;
(e) Agree not to disclose, at any stage of the proceedings, documents or information that the Prosecutor obtains on the condition of confidentiality and solely for the purpose of generating new evidence, unless the provider of the information consents; and
(f) Take necessary measures, or request that necessary measures be taken, to ensure the confidentiality of information, the protection of any person or the preservation of evidence.

[60] Article 100. Rome Statute of the International Criminal Court.

for the application of the ICC Statute.[61] Also, the Regulations of the ICC provide details about the ICC using experts in a trial. Regulation 44 refers to a list of experts accessible at all times to all organs of the Court and to all participants. With an appropriate indication of expertise in the relevant field, the Court may require an expert report for evidence purposes to be presented.[62]

Cases from the ICC show the importance of a satellite imagery expert in interpreting the satellite images provided by the parties and the need to provide explanations to the judges regarding the procedure to produce satellite images that are used as evidence.

3.1.2.1 Situation in the Republic of Kenya

In the case *The Prosecutor v. William Samoei Ruto and Joshua Arap Sang*, the accused were investigated under the charge of crimes against humanity in the context of the 2007–2008 post-election violence in Kenya.[63] The trial started on 10 September 2013 and the case was terminated on 5 April 2016 on the grounds that the majority of the Chamber concluded that the Prosecution had not presented sufficient evidence on which a reasonable Trial Chamber could convict the accused.[64]

However, this case is interesting from the view of the expert analysis of satellite imagery and the request from the Court.[65] The Prosecution proposed three areas of expert evidence, including satellite imagery next to social and political background

---

[61] The Rules of Procedure and Evidence are reproduced from the Official Records of the Assembly of States Parties to the Rome Statute of the International Criminal Court, First session, New York, 3-10 September 2002 (ICC-ASP/1/3 and Corr.1), part II.A. The amendments are reflected via footnotes.

[62]
1. The Registrar shall create and maintain a list of experts accessible at all times to all organs of the Court and to all participants. Experts shall be included on such a list following an appropriate indication of expertise in the relevant field. A person may seek review by the Presidency of a negative decision of the Registrar.
2. The Chamber may direct the joint instruction of an expert by the participants.
3. On receipt of the report prepared by an expert jointly instructed, a participant may apply to the Chamber for leave to instruct a further expert.
4. The Chamber may proprio motu instruct an expert.
5. The Chamber may issue any order as to the subject of an expert report, the number of experts to be instructed, the mode of their instruction, the manner in which their evidence is to be presented and the time limits for the preparation and notification of their report.

[63] International Criminal Court, Case: The Prosecutor v. William Samoei Ruto and Joshua Arap Sang, https://www.icc-cpi.int/Pages/record.aspx?docNo=ICC-01/09-01/11-492-AnxB1, (accessed 24.09.2019).

[64] ICC ICC-01/09-01/11, The Prosecutor v. William Samoei Ruto and Joshua Arap Sang, https://www.icc-cpi.int/CaseInformationSheets/RutoSangEng.pdf, (accessed 24.09.2019).

[65] International Criminal Court, No. ICC-01/09-01/11, Prosecution's report on joint instruction of experts, 31 October 2012, https://www.icc-cpi.int/CourtRecords/CR2012_09279.PDF, (accessed 17.06.2019).

and Post-Traumatic Stress Disorder.[66] The Prosecution proposed Mr. Lars Bromley as a satellite imagery expert to testify in both Kenya cases and provided both Defence teams with his proposed instructions. The satellite imagery expert was asked for the availability of satellite images for the identified locations and times, and was provided with a chart listing locations and time.[67] He was then asked by the ICC to testify on certain matters and provide a written report including his analysis in this case.[68,69]

In the case of *The Prosecutor v. Francis Kirimi Muthaura and Uhuru Muigai Kenyatta*, in advance of trial, the prosecution re-reviewed its evidence against both Muthaura and Kenyatta. Kenyatta was accused in 2010 of five counts of crimes against humanity in the context of the 2007-2008 post-election violence in Kenya. The charges were withdrawn in December 2014.[70] The Prosecution withdrew the charges against Muthaura on the basis that there was no reasonable prospect of conviction if it were to proceed to trial.[71] Also, in this case, the prosecution instructed a Mr. Bromley as its own expert in satellite imagery.[72]

---

[66] Ibid. *supra* note International Criminal Court, No. ICC-01/09-01/11, Prosecution's report on joint instruction of experts. Paragraph 2, 31 October 2012.

[67] Ibid. *supra* note International Criminal Court, No. ICC-01/09-01/11, Prosecution's report on joint instruction of experts, Paragraph 10–12, 31 October 2012.

[68] International Criminal Court, Proposed Instruction for Satellite Imagery Expert, https://www.icc-cpi.int/RelatedRecords/CR2012_09891.PDF, (accessed 17.06.2019).

[69] Within the legal framework of the International Criminal Court and in the cases of the Prosecutor v. William Samoei Ruto and Joshua Arap Sang and the Prosecutor v. Francis Kirimi Muthaura and Uhuru Muigai Kenyatta, you are kindly requested, to the extent to which your expertise permits, to conduct a satellite imagery analysis for certain satellite images based on the Charts annexed to this letter. (…)

You are requested to prepare to testify on and to provide a written report including your analysis describing the following:

(i) The type of physical objects and movements/activity that can be discerned when analyzing the satellite images.
(ii) The status of the physical objects depicted in the satellite images. In particular, whether any of which were subject to acts of destruction or burning on or around the given times. This analysis may require comparing images from different time periods.
(iii) The methodology in preparing the Expert report, including:
    (a) The material and computer software you used;
    (b) How the Satellite Images were stored, and if applicable, the measures taken to ensure the Satellite Images were not adulterated.
    (c) How you analysed the Satellite Images, in particular explaining the scope for and/or limitations in analysing the Satellite Images for the Objects as outlined above.

[70] ICC-01/09-02/11, Case Information Sheet, 13 March 2015, https://www.icc-cpi.int/CaseInformationSheets/kenyattaEng.pdf, (accessed 26.09.2019).

[71] ICC-01/09-02/11, Prosecution notification of withdrawal of the charges against Francis Kirimi Muthaura, 11 March 2013, https://www.icc-cpi.int/CourtRecords/CR2013_01871.PDF, (accessed 26.09.2019).

[72] International Criminal Court, ICC-01/09-02/11, https://www.icc-cpi.int/CourtRecords/CR2013_01425.PDF, (accessed 24.09.2019).

3.1.2.2 Situation in the Democratic Republic of Congo

In the *Ntaganda Case, the Prosecutor v. Bosco Ntaganda* (ICC-01/04-02/06), the trial opened on 2 September 2015 with closing statements taking place on 28–30 August 2018. In the ICC Trial Chamber VI session on 8 July 2019, Delivery of Judgment, Courtroom I, the ICC declared Bosco Ntaganda guilty of war crimes and crimes against humanity.[73] Bosco Ntaganda was the Alleged Deputy Chief of Staff and commander of operations of the Forces Patriotiques pour la Liberation du Congo (FPLC). According to the ICC website, he was charged with thirteen counts of war crimes and five counts of crimes against humanity allegedly committed in 2002 to 2003 in the Ituri District of the DRC.[74]

Referring to the case, the Single Judge acting on behalf of Pre-Trial Chamber II of the ICC, decided on the Prosecutor's request. In the "Redacted Decision on the Prosecution's request pursuant to Regulation 35 to vary the time limit for submission of an updated expert report on satellite image analysis" (03 July 2014/Pre-Trial Chamber II/Decision) it underlined the use of satellite images.[75] Referring to the expert report disclosed to the Defence on 1 November 2013, the Prosecutor highlighted that the report contained a series of satellite images of locations relevant to the charges: twenty five "long-range" and twenty-one "close-range" satellite images. The prosecutor requested from the ICC case judge "to be authorized to disclose an updated version of the expert report with updated titles for the locations in the close-range images and the expert's explanation on the methodology he used when analysing the close-range images".[76] Noting Articles 61 and 67 of the ICC Rome Statute, Rules 121 and Regulation 35 of the Regulations of the ICC, in particular taking into account Regulation 35 (2) referring to variation of time limits in exceptional cases,[77] the case judge accepted the "good cause" shown by the Prosecutor and granted, on an exceptional basis referring to the time limit for evidence, the request to disclose an updated version of the report with more details about the analysis of the satellite imagery.

The transcript of the trial phase from 9 December 2016 contains a series of satellite images and the analysis performed to provide proof of the facts requested by the ICJ.[78]

---

[73] ICC website trial, ICC-01/04-02/06, 08.07.2019, https://www.icc-cpi.int/drc/ntaganda, (accessed 25.06.2019).

[74] ICC-01/04-02/06, Case Information Sheet, July 2019, https://www.icc-cpi.int/CaseInformationSheets/NtagandaEng.pdf, (accessed 27.09.2019).

[75] ICC decision, https://www.icc-cpi.int/Pages/record.aspx?docNo=ICC-01/04-02/06-175-Red, (accessed 25.06.2019).

[76] ICC-01/04-02/06, 3 July 2014, Paragraph 7, https://www.icc-cpi.int/CourtRecords/CR2014_06091.PDF, (accessed 25.06.2019).

[77] The Chamber may extend or reduce a time limit if good cause is shown and, where appropriate, after having given the participants an opportunity to be heard. After the lapse of a time limit, an extension of time may only be granted if the participant seeking the extension can demonstrate that he or she was unable to file the application within the time limit for reasons outside his or her control.

[78] ICC-01/04-02/06-T-175-Red-ENG, Trial Transcript, 9 December 2016, https://www.icc-cpi.int/Pages/record.aspx?docNo=ICC-01/04-02/06-T-175-Red-ENG, (accessed 24.09.2019).

The transcript contained information that the new satellite images and the analysis had been conducted and presented to the ICC.[79] The report provided to the ICC was of a technical nature and the ICC proceeded to the testimony of Mr. Bromley, called by the Prosecution, to testify as an expert and to assist the ICC in finding the truth. The Judge of the case asked the expert to tell the truth and provide clarifications to help the ICC to find the truth: *"We want you to tell the truth about this main issue. We want you to only testify within the sphere of your expertise. You are before us an expert of satellite imagery and will be testifying in relation to the report you submitted to the Prosecution."*[80]

Mr. Lars Eric Bromley born in Quito, Ecuador was the expert in the matter, providing a satellite imagery analysis of a number of locations of interest in the form of coordinates provided either by the ICC using a global positioning device or by referencing maps from various sources.[81] In particular, the expert had to indicate to the Court the settlement locations, including in the DRC, to acquire relevant satellite images using verified coordinates for locations of interest, analyse the images and "document any potential destruction, burning or damage to structures". The Court asked the expert if, based on the analysis of satellite images, he found possible destruction, burning or damage to structures.

Information about Mr. Bromley activity was provided in the transcripts.[82] Mr. Bromley was responsible for the analysis of satellite imagery and geospatial data. He worked at the American Association of the Advancement of Science and the UN office of the High Commissioner of Human Rights investigations of human rights violations in Libya. He had published extensively, including in 2010 in the International Journal of Remote Sensing "Relating Violence to MODIS"—Fire Detections in Darfur Sudan, and in 2009, Eye in the Sky: Monitoring Human Rights Abuses using geospatial technology, as well as in 2007, Towards a Human Rights Observing system.

According to the explanations to the prosecutor, starting April 2014, Bromley analysed satellite imagery and geospatial data to develop expert reports in the case of the "Prosecutor and Ruto and Sang", as well as the case of "Prosecutor and Kenyatta". And he had also testified as an expert witness in the case against Mr. Ruto and Mr. Sang. Bromley further explained his expertise in analysing satellite imagery and daily activity at UNOSAT. The testimony provided information about the request from different UN organizations to develop analytical products to verify sites of destruction, displacement or assessing the risk for refugee camps.

[79]Ibid. *supra* note ICC-01/04-02/06-T-175-Red-ENG, pp. 66 of the transcripts, 9 December 2016, https://www.icc-cpi.int/Transcripts/CR2016_26069.PDF, (accessed 24.09.2019).

[80]Ibid. *supra* note ICC-01/04-02/06-T-175-Red-ENG, pp. 69 of the transcripts, 9 December 2016, https://www.icc-cpi.int/Transcripts/CR2016_26069.PDF, (accessed 24.09.2019).

[81]Ibid. *supra* note ICC-01/04-02/06-T-175-Red-ENG, pp. 70, of the transcripts, 9 December 2016, https://www.icc-cpi.int/Transcripts/CR2016_26069.PDF, (accessed 24.09.2019).

[82]Ibid. *supra* note ICC-01/04-02/06-T-175-Red-ENG, Trial Transcript, pp. 70 of the transcripts, 9 December 2016, https://www.icc-cpi.int/Transcripts/CR2016_26069.PDF, (accessed 24.09.2019).

> What we do at UNOSAT is we are essentially an on-call satellite imagery analysis centre available to the UN system and other organizations. We generally receive requests from UN partners to understand issues that are going on in certain parts of the world that may be inaccessible or too dangerous to visit. This could include populations of displaced persons or areas of combat or refugee camps or similar issues where we essentially locate the event in question or the town or city in question. We acquire satellite imagery once we know what is available and we develop analytical products in response to the requests that we receive. So, we are very often verifying reports of destruction, verifying reports of displacement, verifying the existence of remote displaced populations, to more esoteric things, like understanding changes in agricultural production, understanding the potential for flood risk for refugee camps and a variety of related issues.[83]

The CV of Bromley was admitted as evidence of his qualifications. Regarding the acquisition of satellite images and the methodology for the actual analysis of those images, the Prosecutor underlined that the data provided by Bromley and analyzed by the Court was found in the preliminary report from the expert dated 17 April 2015 identified by ERN DRC-OTP-2084-0443 and the final report dated 14 November 2016 identified by ERN DRC-OTP-2099-0166.

The Court was interested in how the expert verified the geocoordinates used to obtain satellite imagery but also in the process to obtain the geocoordinates for specific locations of interest that Bromley used for the satellite imagery analysis. The expert explained the process step by step, highlighting that they were very often trying to locate remote areas in remote towns: "I used two different publicly available data sets. The first is the GEONet Names Server, GNS, which comes from the U.S. government; the second one is called RGC, which comes from the DRC government."[84] Very interesting for the process was that each satellite had its own orthorectification characteristics.[85] Furthermore, the U.S. government database, the GNS, is produced by a U.S. Defense Department organization called the National Geospatial Intelligence Agency. It is made publicly available for any type of geospatial cartography mapping application where it includes probably millions of points for the entire world. The expert underlined that it is practically a nationwide cadastral dataset, a dataset explaining where everything is in a particular country.

Regarding the acquisition of the satellite images, it was stated that satellite imagery was available for ICC Areas of Interest in the DRC, namely ERN DRC-OTP-2099-0284. A question from the prosecution referred to the acquisition of the satellite images. Bromley stated that the vendors were European Space Imaging and then a Norwegian company called Kongsberg Satellite Applications Systems. Bromley further mentioned that the original imagery had been collected by two U.S. companies, one called DIGITAL GLOBE, which operated the QUICKBIRD satellite, and the second company at the time of the image acquisition was called Space Imaging. They operated the IKONOS satellite space imagining, later becoming GOI, which

[83] Ibid. *supra* note ICC-01/04-02/06-T-175-Red-ENG, Trial Transcript, pp. 75 of the transcripts, 9 December 2016, https://www.icc-cpi.int/Transcripts/CR2016_26069.PDF, (accessed 24.09.2019).

[84] Ibid. *supra* note ICC-01/04-02/06-T-175-Red-ENG, Trial Transcript, pp. 79 of the transcripts, 9 December 2016, https://www.icc-cpi.int/Transcripts/CR2016_26069.PDF, (accessed 24.09.2019).

[85] Ibid. *supra* note ICC-01/04-02/06-T-175-Red-ENG, Trial Transcript, pp. 81 of the transcripts, 9 December 2016, https://www.icc-cpi.int/Transcripts/CR2016_26069.PDF, (accessed 24.09.2019).

was later bought by DIGITAL GLOBE. Bromley said that in practice, at the time he purchased the imagery, it was all bought from the original DIGITAL GLOBE through the two vendors EUSI and KSAT.[86]

Referring to the availability of satellite images, Bromley gave an overview of the procedure. What was underlined was that satellite imagery is practically collected based on client orders. No satellite currently collects images from all places on the globe, and there is no "Big Brother" type activity all around the Earth but rather strategic gathering over certain areas.

> The two satellites that we used, the IKONOS satellite and the QuickBird satellite, had been operating, I think I have it since, since about 2000 and 2002. From their start of operation, their collecting imagery, both based on client orders, so anyone in this room can essentially order a satellite image to be collected through these companies. The companies themselves are also essentially doing what you can think of as strategic image gathering, so if someone is not ordering an image to be collected over a certain area, these companies are gathering imagery on their own.[87]

Asking for a clarification, the prosecution asked the expert to clarify whether there would always be a satellite image every day in every location.

> For high value economic areas like something like New York City or The Hague or something, you will see a lot of imagery acquired from the start of the satellite's operation. For rural areas like Eastern DR Congo, it is really going to be a much more haphazard set of imagery collected. So if you have a certain location, you can search for imagery for that location and you may find one image back in the years that we are interested in, you may find two images, or you may find zero. Again, it is going to go back to what the companies were actually doing at that time, what their priorities and clients were will dictate essentially where they're collecting imagery. Now we are talking about the high-resolution imagery which is useful for this analysis where we can actually see houses and vehicles and stuff like that. There are, of course, weather satellites and other satellites which are gathering imagery for the entire world all the time. However, those satellites don't provide the level of detail where you are going to see things like individual houses or roads or cars.[88]

The prosecutor also asked the expert to explain the dates of these images, and how the dates were recorded for each of the images that were acquired.

> Every time a satellite image is acquired, it is—it is not just a photograph of the ground. It is basically a fairly advanced data product. So you have the image itself, which is the image of the ground that you'll see in many of the figures in my report. At the same time, it is basically timestamping the image acquisition, literally down to the picosecond. Reference to Greenwich Mean Time or Zulu Time, as it is called in the industry. Then it is collecting, it is also recording a lot of other information at the same time: The satellite position, the sun position, a whole mass of additional information which most of us don't ever need to get into. But the image itself is, is timestamped at the moment of acquisition.[89]

---

[86] Ibid. *supra* note ICC-01/04-02/06-T-175-Red-ENG, Trial Transcript, pp. 87 of the transcripts, 9 December 2016, https://www.icc-cpi.int/Transcripts/CR2016_26069.PDF, (accessed 24.09.2019).

[87] Ibid. *supra* note ICC-01/04-02/06-T-175-Red-ENG, Trial Transcript, pp. 87-88 of the transcripts, 9 December 2016, https://www.icc-cpi.int/Transcripts/CR2016_26069.PDF, (accessed 24.09.2019).

[88] Ibid. *supra* note ICC-01/04-02/06-T-175-Red-ENG, Trial Transcript, pp. 88 of the transcripts, 9 December 2016, https://www.icc-cpi.int/Transcripts/CR2016_26069.PDF, (accessed 24.09.2019).

[89] Ibid. *supra* note ICC-01/04-02/06-T-175-Red-ENG, Trial Transcript, pp. 89 of the transcripts, 9 December 2016, https://www.icc-cpi.int/Transcripts/CR2016_26069.PDF, (accessed 24.09.2019).

The Prosecutor acknowledged that in terms of the satellite images obtained by the expert, the ICC has been provided with the metadata referred to in the expert report. The Prosecutor requested the expert to confirm that has provided analysis of the 25 satellite images from 2013 disclosed to Defense, five satellite images disclosed on 17 April 2015 and finally two satellite images from 2 December 2016. The expert expressed its consent for the admission of these satellite images as evidence in the case. The presiding judge admitted this evidence as a further Prosecution exhibit.[90] The prosecutor asked the expert about the actual analysis of the satellite imagery. His first question was the impact on the quality of the satellite images themselves, the types of problem in terms of quality.

> So each satellite has a, what we call a maximum resolution of what it can image. So the QuickBird satellite, for example, is roughly imaging things 70 centimetres across. That describes how it will image the ground if it is directly overhead. The satellite itself can essentially swivel a little bit and look off to the side, so the more it looks off to the side the more the image quality will be degraded. In addition, there is of course cloud cover. Especially in an area like DR Congo, clouds are persuasive (sic), clouds will generally block out an area. If they are, you know, large white fluffy clouds they are going to completely block an area. Sometimes you can see through the hazy clouds and see what is on the ground, but the haze is certainly going to degrade image quality as well. Finally, there is other factors like a ground level mist or a fog which may not be very apparent, but you will just notice that the image that you are looking at is lower quality than what you are used to. There is one other type of clouds, the very high atmospheric cirrus clouds which are essentially a, kind of a thin cloud high in the atmosphere, they can also degrade the image quality. Before you order an image, you can look at a low-quality sample of that image to understand where the cloud cover is and try and minimize your purchase of clouds, basically.[91]

The prosecutor further insisted that the expert state if any of the satellite imagery from the location coordinates that were provided was not purchased due to problems of clarity such as clouds covering the location. The expert admitted that for several locations the imagery was a hundred per cent cloudy over the location of interest and there was no point in buying such image.[92]

Referring strictly to the analytical procedure, the prosecutor asked the expert to provide a particular analysis to document and identify potential structures or potential destruction or damage and to provide clarifications.

> Once you acquire the imagery, it's relatively straightforward. If you have two images collected before and after the event in question, you are generally zooming into an image to a level where you can see the details like small houses, small paths and roads and things like that. And then you are simply comparing the two images together. You want the image before the event and then the image after the event and you are then documenting the changes in structures that you see between those two images if you see them. So if you see a house in the image collected before an event and the house is not there in the image after the event,

[90] Ibid. *supra* note ICC-01/04-02/06-T-175-Red-ENG, Trial Transcript, pp. 91 of the transcripts, 9 December 2016, https://www.icc-cpi.int/Transcripts/CR2016_26069.PDF, (accessed 24.09.2019).

[91] Ibid. *supra* note ICC-01/04-02/06-T-175-Red-ENG, Trial Transcript, pp. 92 of the transcripts, 9 December 2016, https://www.icc-cpi.int/Transcripts/CR2016_26069.PDF, (accessed 24.09.2019).

[92] Ibid. *supra* note ICC-01/04-02/06-T-175-Red-ENG, Trial Transcript, pp. 93 of the transcripts, 9 December 2016, https://www.icc-cpi.int/Transcripts/CR2016_26069.PDF, (accessed 24.09.2019).

> you can be very certain that the house has basically disappeared and then you are marking that with a coordinate point that is stored in a database. In locations where you only have one image collected after the event, which was common in this analysis, you are then looking for characteristics in the imagery where you would expect to see structures but you don't see structures. So as you analyse an area the analyst does gain some familiarity with how the geography is, where houses are positioned next to roads or next to paths, what a homestead might look like with multiple structures, and then you are identifying those locations where, in essence, you think there should be those structures but you are not seeing them. Obviously, when you use the two images, the before and the after, the results are, are much more certain than when you use the one image collected after. With the one image after, you will see in the report that it says it's is a speculative method. Within the speculation there are some areas which are more certain and some areas which are less certain, but I didn't have time to document that range in the report.[93]

The prosecutor sought clarification about the language that the expert used for the degree of certainty. The request focused on the likelihood of describing the events as happened and identifying potential challenges in collecting the satellite imagery.

> There is a whole, or a small kind of vocabulary which we as image analysts are known to use. We are always cognisant of the fact that we are not there on the ground, we are looking at a remote view of the area, we are not seeing all the details that you would see if you were standing there on the ground. So if I'm doing an analysis in an area and I locate a hundred structures which I think are destroyed, you will see me use words like "possible" or "likely" and that, at a basic level, just reflects the fact that I was not there when whatever happened on the ground happened. I cannot, you know, say for certain that a destructive event took place versus some sort of, you know, town redevelopment project, for example. In addition, if I find a hundred structures that had been destroyed in the imagery, if we were able to go back in time and actually visit that location, you are certainly going to find, you know, five or 10 per cent of those structures where I was wrong. Perhaps a tree branch blew over the house at the time of image acquisition. So over wide areas words like "possible" and "likely" are essentially used to, you know, build in that error margin that is just basically an artefact of the fact that we are using satellite imagery versus actually standing there on the ground witnessing in full, full detail what is going on.[94]

Further, the prosecutor asked about the meaning of the term "speculative" and if it was ordinarily used by satellite imagery analysts. The expert stated that such term appears only when using a single image and such image is only collected after the event occurred.

> The main problem there is I cannot say that structures disappeared if I do not know that there were structures there to begin with. Now sometimes we are seeing the remains of structures, you are seeing burned, charred, you know, structural remains and I would still use the word speculative there simply because I do not know what that location looked like. However, the burned charred remains, to me, are something that are, you know, more certain than other areas where I am seeing a cleared area which looks like it should have houses in it along a road but I see no houses.[95]

[93] Ibid. *supra* note ICC-01/04-02/06-T-175-Red-ENG, Trial Transcript, pp. 94 of the transcripts, 09 December 2016, https://www.icc-cpi.int/Transcripts/CR2016_26069.PDF, (accessed 24.09.2019).

[94] Ibid. *supra* note ICC-01/04-02/06-T-175-Red-ENG, Trial Transcript, pp. 95 of the transcripts, 9 December 2016, https://www.icc-cpi.int/Transcripts/CR2016_26069.PDF, (accessed 24.09.2019).

[95] Ibid. *supra* note ICC-01/04-02/06-T-175-Red-ENG, Trial Transcript, pp. 95 of the transcripts, 9 December 2016, https://www.icc-cpi.int/Transcripts/CR2016_26069.PDF, (accessed 24.09.2019).

Referring to the type of software package used for the review of the satellite imagery, the expert underlined that the general family of the software is called "Geographic Information Systems" and the specific type used is called "ArcMap". The ArcMap GIS was used for the analysis of the satellite imagery and the expert reports provided to the Court. Each image provided to the report was analysed multiple times:

> For each image it was definitely gone through, each image was analysed at least twice and I mean some of them at least half a dozen, maybe more times after that. By the time it made it into this report, the final report, each image or pair of images had been gone through multiple times.[96]

The expert provided a PowerPoint presentation stating that the software ArcMap GIS analysis is a very common software system, used in many sectors for different purposes, especially the defense related sector and also commercial sectors. Also, Bromley stated that when you get into GIS work "you go from seeing a basic map to seeing a dataset". The global borders, seen as what is called layers, can essentially turn on and off. This software had been used for almost 20 years now, and was considered as common. The presentation contained analysis of pictures and a visual presentation of the procedure. At the request of the prosecution, the expert further explained the meaning of the data such as the red, green, and blue bands.

> When a satellite like these capture an image, they are not just taking a simple photograph. These are sensors that are detecting the reflected electromagnetic energy which originates from the sun, hits the surface of the earth and bounces back up into the sensor. These particular satellites are collecting this electromagnetic energy in the red, green and blue bands, which are what we would call the visual spectrum, and that's the same thing that we see when we look at these images. This is a natural colour image, this is essentially what you would see if you were flying overhead in an aircraft. These satellites are also collecting a fourth band of imagery, the near infrared which our eyes don't see but can be detected by a sensor and by various insects and animals as well, and occasionally we use that fourth band for certain parts of the analysis.[97]

The expert further referred to the fourth band of the near infrared for the process of analysis, providing explanations why the colour red was used, to which he replied that it was used for contrast, to determine the differences on the images.[98]

About the methodology used, the expert provided a comparison of different images:

> In January I have marked multiple rooftops with the red arrows. These are square structures. You can actually see the peak in the roof like you would on any peaked roof structure for several of these houses. If we go over to 22 May 2003, drawing your eye down to the bottom especially, all the structures that are visible along the bottom have essentially disappeared. You are no longer seeing these nice sharp, square, square structure lines, you are seeing more

---

[96] Ibid. *supra* note ICC-01/04-02/06-T-175-Red-ENG, Trial Transcript, pp. 96 of the transcripts, 9 December 2016, https://www.icc-cpi.int/Transcripts/CR2016_26069.PDF, (accessed 24.09.2019).

[97] Ibid. *supra* note ICC-01/04-02/06-T-175-Red-ENG, Trial Transcript, pp. 103-104 of the transcripts, 9 December 2016, https://www.icc-cpi.int/Transcripts/CR2016_26069.PDF, (24.09.2019).

[98] Ibid. *supra* note ICC-01/04-02/06-T-175-Red-ENG, Trial Transcript, pp. 103-104 of the transcripts, 9 December 2016, https://www.icc-cpi.int/Transcripts/CR2016_26069.PDF, (24.09.2019).

> irregular patches of brown which are basically the ground, indicating that the structure has disappeared.[99]

The defense submitted a PowerPoint on September 2015 but did not refer to the satellite images.[100]

3.1.2.3 Blog: International Criminal Court

Satellite imagery as evidence of international crimes was discussed in a blog post part of the International Justice Monitor's Technology for Truth series. It was acknowledged that satellite imagery has the particular ability to provide "*unique, otherwise unavailable evidence of alleged mass atrocity crimes. Because these events often occur in non-permissive environment, over large geographic areas, and across long- and multiple-time frames, high resolution satellite imagery, in particular, has value for international tribunals investigating these crimes*".[101] It was underlined that satellite imagery was first used in international criminal proceedings during the Srebrenica trials at the International Criminal Tribunal for the Former Yugoslavia. It was stated that:

> Satellites, in the simplest sense, capture light reflecting from the surface of the Earth. Once the light is captured and digitized, it is processed into an image. The images can be adjusted and processed to isolate, amplify, or suppress specific wavelengths of light that can provide the analyst with important information beyond what the human eye can perceive. Through imagery interpretation and analysis, visual information can be used to cross-corroborate complementary non-imagery information—witness testimony, reports and newspaper articles, or information gathered by crowdsourcing platforms. The result of this correlation of data streams is later published as a finding, which can be offered as evidence.

In general, the phenomena that can be detected with the help of satellite imagery include: "cratering consistent with bombardment; the destruction of civilian dwellings; targeting of humanitarian and other forms of internationally protected infrastructure; the locations, size, and composition of displaced persons camps; excavations consistent with mass graves; and other probative phenomena". Satellite evidence can provide critical crime-specific and circumstantial evidence. Most importantly, "It should not be assumed, however, that a specific fact in issue can be proven on the basis of satellite imagery analyses alone. In many instances corroborative evidence is needed, which is often integrated into the analysis". For this reason, the main challenges for the use of satellite imagery as evidence before the ICC were identified as follows:

---

[99] Ibid. *supra* note ICC-01/04-02/06-T-175-Red-ENG, pp. 109 of the transcripts, 9 December 2016, https://www.icc-cpi.int/Transcripts/CR2016_26069.PDF, (24.09.2019).

[100] ICC, Annex 4 to the Registration into the Record of Material Presented during the hearings of the 2nd and 3rd September 2015 https://www.icc-cpi.int/Pages/record.aspx?docNo=ICC-01/04-02/06-886-Anx4, (accessed 24.09.2019).

[101] Kroker, P., Satellite imagery as Evidence for International Crimes" in International Justice Monitor, 2015, https://www.ijmonitor.org/2015/04/satellite-imagery-as-evidence-for-international-crimes/, (accessed 1.07.2019).

1. The limited availability of appropriate imagery data;
2. Restricted budgets and cognitive bias of external organizations conducting the analysis;
3. No accepted forensic standards and methodologies, and
4. A lack of familiarity of judges with this kind of evidence.

Among the reasons for the limited availability were meteorological phenomena that obstruct the image of the location, such as clouds. This was stated by the satellite imagery expert at the ICC case. However, among other elements were also government restrictions on collecting imagery of a certain area, and the high financial cost of obtaining imagery.

High-resolution imagery that could be made available is mainly collected by commercial providers that only collect data when it is of current of potential future value. It was mentioned that "*Whether imagery of a specific situation is pertinent for a criminal investigation can often only be known months, sometimes years later. Hence commercial providers have a low economic incentive to routinely collect areas where mass atrocities are allegedly occurring*". Once you pass the moment to collect that image, it is an insurmountable barrier, the data is lost.

Reliance on External Capacity Is Another Challenge for the International Criminal Court

It was put forward that the ICC does not possess the capability yet to perform the analysis work with an in-house expert which is then often performed by third party analysts who are not affiliated with a specific court or prosecutor office. It was mentioned that a number of organizations perform data analysis for the court, such as: Human Rights Watch, UNOSAT, Amnesty International, the Signal Program at the Harvard Humanitarian Initiative, and AAAS. The financial restrictions were also stated as a reason for the limited number of contributions. In particular, analysis of images is complex and the data on which the analysis is made needs to come from a trusted source, to ensure that the ICC takes into account reasonable forms of evidence.

No Accepted Forensic Standard and Methodologies

The lack of an accepted forensic standard has emerged as a challenge. The growing number of satellite data analyses of human rights violations are performed by different experts, using a number of programs, which could also be the same. However, analyses use different colours, resolution, and zoom. Analysts currently operate according to their own best practices and experience but it has been argued that they should operate under an accepted forensic standard for remote sensing. Such a solution should be welcomed by the ICC so that, under standardized training, judges could more conveniently understand how satellite images become evidence.

Also, another challenge is general unfamiliarity of judges with geospatial evidence and the need to discuss the methodology and have clarifications from the experts. Described as a complex technical and methodological processes based on the division of labour, satellite imagery analysis needs to be able to become evidence and be given appropriate weight in finding the truth. In ICC cases the current practice is the presenting of images and findings in pdf-printouts on in-court computer screens. Several challenges here require attention and solutions:

- The accuracy, objectivity and authenticity of the data must be proven during the court procedure. For satellite data to be admissible, it needs to be reliable.
- The importance of the chain of custody of the data along the process has been noted. The court needs to assess the number of systems and individuals who contributed to the analysis and what the result of this analysis is. Such a solution requires the development of standardized methods that include: "data acquisition, storage, interpretation, integration of cross-reference data, and visualization for the specific purposes of international criminal investigations".[102] The advancement of satellite imagery as evidence in courts must address the standardization of image processing and data interpretation together with the authentication or certification of the final result.
- The practical circumstances of the collection and analysis of satellite images must also be addressed. Such circumstances may refer to the cost of acquisition, the companies that provide the data, the areas for which the data is collected, the period before the event which is to be compared with other data during and after the event.
- There is also an urgent need to familiarize lawyers and especially judges with this process, to give them confidence in assessing such evidence and the means to evaluate its reliability and probative value.[103]

If the above challenges can be overcome and the evidentiary potential of satellite-based evidence is fully exploited, "prosecutions at the ICC could rely on a unique form of evidence, which would enhance the quality of international prosecutions and contribute to the fight against impunity of mass perpetrators".[104]

Under the Harvard Humanitarian Initiative, in December 2014 Dr. Patrick Kroker published the "Emerging Issues Facing the Use of Remote Sensing Evidence for International Criminal Justice".[105] The declared purpose of this study was to: (i)

[102] Ibid. *supra* note Kroker, P., Satellite imagery as Evidence for International Crimes" in International Justice Monitor, 2015, https://www.ijmonitor.org/2015/04/satellite-imagery-as-evidence-for-international-crimes/, (accessed 2.07.2019).

[103] Ibid. *supra* note Kroker, P., Satellite imagery as Evidence for International Crimes" in International Justice Monitor, 2015, https://www.ijmonitor.org/2015/04/satellite-imagery-as-evidence-for-international-crimes/, (accessed 2.07.2019).

[104] Ibid. *supra* note Kroker, P., Satellite imagery as Evidence for International Crimes" in International Justice Monitor, 2015, https://www.ijmonitor.org/2015/04/satellite-imagery-as-evidence-for-international-crimes/, (accessed 2.07.2019).

[105] Kroker, P., Emerging Issues Facing the Use of Remote Sensing Evidence for International Criminal Justice, December 2014, http://hhi.harvard.edu/publications/emerging-issues-facing-use-remote-sensing-evidence-international-criminal-justice-0, (accessed 24.09.2019).

illustrate the nature of Remote Sensing (RS) analysis and its evidentiary potential and limitations; (ii) identify the types of crimes that influence its limited use in court; and (iii) explore steps and strategies for overcoming these challenges.

In relation to the Nature of Remote Sensing Analysis, Kroker mentioned that the majority of remote sensing analyses of alleged crimes are performed by third party analysts who are not affiliated with a specific court or prosecutor's office. Kroker underlined that before and during analysis of the remote sensing data, the original satellite image is manipulated and adjusted by the expert, providing the specific steps for the image interpretation. Hence, remote sensing analyses are processed by a number of systems and individuals before submitting the evidence to court. The author divided remote sensing analysis into evidentiary value and limitations of remote sensing. Regarding the benefits of remote sensing as evidence for the court, Kroker mentioned that satellite imagery can provide critical crime-based and circumstantial evidence. As limitations, Kroker underlined: "(1) *It is unlikely that a specific fact in issue can be proven on the basis of RS analyses alone. In most instances, corroborating evidence is needed. This evidence can sometimes be integrated into the analysis, if it is available.* (2) *The ability to provide for direct and (specifically useful) linkage evidence is limited.* (3) *Only a relatively small number of elements of international crimes can be shown, typically including the widespread or systematic character of an attack.*"[106]

Referring to the aspects that influence the use of remote sensing as evidence, Kroker highlighted: (i) the reliance on third party investigators; (ii) limited availability of appropriate high-resolution imagery; (iii) cognitive bias and different focus; (iv) lack of forensic criminalistics; (v) presentation and evaluation of evidence; (vi) legal aspects affecting the use of remote sensing analysis as evidence; (vii) accuracy; (viii) objectivity; (ix) authenticity and (x) chain of custody. The paper also contained recommendations. Among the most important elements, Kroker emphasized the need to define standards for admission and evaluation of remote sensing evidence. He also underlined the need for development of best practice standards among analysts and software and hardware platforms to facilitate the demonstration of the authenticity of the data. Finally, there is a need to familiarize the lawyers, prosecutors and judges with the interpretation of remote sensing evidence.[107]

[106] Ibid. *supra* note Kroker, P., Emerging Issues Facing the Use of Remote Sensing Evidence for International Criminal Justice, December 2014, http://hhi.harvard.edu/publications/emerging-issues-facing-use-remote-sensing-evidence-international-criminal-justice-0, (accessed 24.09.2019).

[107] Ibid. *supra* note Kroker, P., Emerging Issues Facing the Use of Remote Sensing Evidence for International Criminal Justice, December 2014, http://hhi.harvard.edu/publications/emerging-issues-facing-use-remote-sensing-evidence-international-criminal-justice-0, (accessed 24.09.2019).

3.1.3 *International Tribunal for the Law of the Sea*

The International Tribunal for the Law of the Sea (ITLOS, Tribunal) is an independent judicial body established by the UN Convention on the Law of the Sea (UNCLOS), located in Hamburg, Germany. Its function is to adjudicate disputes arising out of the interpretation and application of UNCLOS. Annex VI "Statute of the International Tribunal for the Law of the Sea" (Statute) provides that the International Tribunal for the Law of the Sea (ITLOS) shall function in accordance with the provisions and Statute of UNCLOS.[108] According to the provisions of Article 21 of the Statute, ITLOS has jurisdiction over any dispute concerning the interpretation or application of UNCLOS and all matters specifically provided for in any other agreement that confers jurisdiction to ITLOS. Under Article 187 "Jurisdiction of the Seabed Disputes Chamber" disputes between states may be submitted to a special chamber of ITLOS, to an ad hoc chamber of the Seabed Disputes Chamber, or to binding commercial arbitration. Some ITLOS cases are of relevance to the evidentiary value of satellite remote sensing, as follows.

3.1.3.1 Delimitation of Maritime Boundary Between Ghana and Côte d'Ivoire

The dispute in this case concerned delimitation of the maritime boundary between Ghana and Côte d'Ivoire in the Atlantic Ocean. ITLOS published on its website the Judgement of 23 September 2017 and provided the reason for its decision.[109] (Case No. 23). Related to satellite imagery, Ghana argued that "Côte d'Ivoire used two different methods to chart the coast on either side of the land boundary terminus, by applying ground survey data only for Côte d'Ivoire's coast. Ghana argued that "satellite-derived bathymetry … is an inappropriate means of constructing a low water line in cases, like this one, where the waters display very high turbidity and breaking waves".[110] Côte d'Ivoire contended that "it is not a valid alternative solution to use the coastline drawn by EOMAP, a company commissioned by Ghana". It further stated that the work carried out by EOMAP was unsatisfactory on several grounds: no in situ survey was conducted; the satellite images covered a very short period and were chosen arbitrarily by that company; and the scale of those images was

[108] UN General Assembly, Convention on the Law of the Sea, 10 December 1982 https://www.un.org/depts/los/convention_agreements/texts/unclos/unclos_e.pdf, (accessed 24.09.2019).

[109] International Tribunal for the Law of the Sea, Dispute concerning delimitation of the Maritime Boundary Between Ghana and Côte d'Ivoire in the Atlantic Ocean, 23 September 2017, https://www.itlos.org/fileadmin/itlos/documents/cases/case_no.23_merits/C23_Judgment_23.09.2017_corr.pdf, (accessed 24.09.2019).

[110] Ibid. *supra* note International Tribunal for the Law of the Sea, Dispute concerning delimitation of the Maritime Boundary Between Ghana and Côte d'Ivoire in the Atlantic Ocean, 23 September 2017, Paragraph 331.

not precise enough to produce reliable results on such a small segment of coastline.[111] When referring to the base points, and to ensure they were accurate and able to reflect the coastal geography of the states, Côte d'Ivoire published new, highly accurate, charts prepared on the basis of topographical surveys of the entire Côte d'Ivoire coast at the end of 2014 and on recent high-resolution satellite images.[112] The charts are entitled 001AEM and 002AEM.

In its judgment, the Special Chamber referred to the images and the dispute between Ghana and Côte d'Ivoire and discussed their relevance: The Special Chamber acknowledged that chart 001AEM was of more recent origin than chart BA 1383. However, the Special Chamber underlined that was not convinced by the arguments advanced by Côte d'Ivoire in favour of chart 001AEM. ITLOS did not question the fact that this chart was prepared "on the basis of topographical surveys of the entire Côte d'Ivoire coast at the end of 2014 and may, therefore, reflect the most recent data concerning that coast. The judges underlined what was relevant for the case, particularly the fact that no similar action had been taken for Ghana. The court stated in its judgment that it was evident from the facts advanced by Côte d'Ivoire that no such topographical survey had been undertaken on the coast of Ghana. "Instead, chart 001AEM relied, as far as the coast of Ghana is concerned, on recent high-resolution satellite images whose reliability is questioned by Ghana. It is not for the Special Chamber to decide whether the satellite-derived bathymetry method used in respect of the coast of Ghana was appropriate and leads to reliable results. Instead, it is of relevance for the Special Chamber that different methods were employed for the survey of the Ivorian and Ghanaian coasts. The Special Chamber agrees with Côte d'Ivoire that a more recently prepared chart is preferable in principle but takes the view that it is essential that the same methodology be used for the two coasts in question".[113]

The ITLOS noted in its judgment that charts BA 1383 and SHOM 7786 were used by both parties until at least 2014, which meant that they both agreed on the data. It was important for the Court to mention that the practice was indicative of the parties' common confidence in the reliability of these charts. For this reason, the Special Chamber "will use charts BA 1383/SHOM 7786 as a basis for its considerations and

[111] Ibid. *supra* note International Tribunal for the Law of the Sea, Dispute concerning delimitation of the Maritime Boundary Between Ghana and Côte d'Ivoire in the Atlantic Ocean, 23 September 2017, Paragraph 333.

[112] Ibid. *supra* note International Tribunal for the Law of the Sea, Dispute concerning delimitation of the Maritime Boundary Between Ghana and Côte d'Ivoire in the Atlantic Ocean, 23 September 2017, Paragraph 336.

[113] Ibid. *supra* note International Tribunal for the Law of the Sea, Dispute concerning delimitation of the Maritime Boundary Between Ghana and Côte d'Ivoire in the Atlantic Ocean, 23 September 2017, Paragraph 341. https://www.itlos.org/fileadmin/itlos/documents/cases/case_no.23_merits/C23_Judgment_23.09.2017_corr.pdf, (accessed 24.09.2019).

for its decision concerning the delimitation of the territorial seas, the EEZ and the continental shelves (within and beyond 200 nm) of the two Parties".[114]

In the Special Chamber's view, the disagreement between the parties concerning the appropriate base points "stems in part from disagreements as to which chart is appropriate, as to where to place base points, and as to the fact that only a few base points can be identified on Jomoro". The Special Chamber underlined in its judgement that, while coastal states were entitled to determine base points for the purpose of delimitation, it was under no obligation to accept the base points identified by either of them. This was the reason why the Court might select its own base points on the basis of the geographic particularities of the coast under consideration (*Delimitation of the maritime boundary in the Bay of Bengal (Bangladesh/Myanmar), Judgment, ITLOS Reports 2012*, p. 4, at p. 72, para. 264, quoting the Judgment of the ICJ in *Maritime Delimitation in the Black Sea (Romania* v. *Ukraine), I.C.J. Reports 2009*, p. 61, at p. 108, para. 137).[115]

Finally, the Special Chamber reiterated that it had decided to use chart BA 1383 (see para. 343 in the Judgement) and based its decision on these data. Having assessed the base points advanced by the parties on the basis of this chart, the Special Chamber concluded that for various reasons the base points suggested by the parties were not appropriate.[116]

3.1.3.2 Land Reclamation by Singapore in and around the Straits of Johor (Malaysia v. Singapore)

In a case concerning the dispute over land reclamation activities in and around the straits of Johor (Malaysia v. Singapore), in September 2003 Malaysia requested ITLOS to prescribe provisional measures against Singapore over the effect of Singapore's land reclamation activities upon Malaysia's rights in and around the Straits of Johor.[117] Malaysia claimed that Singapore's action in land reclamation around Pulau Tekong and Tuas was causing damage to the marine environment and serious prejudice to the rights of Malaysia. Malaysia asked that Singapore suspend all land reclamation activities in the vicinity of the maritime boundary between the two states

[114] Ibid. *supra* note International Tribunal for the Law of the Sea, Dispute concerning delimitation of the Maritime Boundary Between Ghana and Côte d'Ivoire in the Atlantic Ocean, 23 September 2017, Paragraph 343.

[115] Ibid. *supra* note International Tribunal for the Law of the Sea, Dispute concerning delimitation of the Maritime Boundary Between Ghana and Côte d'Ivoire in the Atlantic Ocean, 23 September 2017, Paragraph 393.

[116] Ibid. *supra* note International Tribunal for the Law of the Sea, Dispute concerning delimitation of the Maritime Boundary Between Ghana and Côte d'Ivoire in the Atlantic Ocean, 23 September 2017, Paragraph 394.

[117] ITLOS, Request for Provisional Measures, 4 September 2003, https://www.itlos.org/fileadmin/itlos/documents/cases/case_no_12/request_malaysia_eng.1.pdf, (accessed 27.09.2019).

or areas claimed as territorial waters and provide Malaysia with full information as to current and projected works.[118]

According to the Statement of Ms. Cheong,[119] the Advocate of Singapore used satellite images before reclamation to prove the land-based activities affected the quality of water.

> I would like to show you some satellite photographs taken before reclamation commenced at Pulau Tekong. These photographs would show that land-based activities have long affected the quality of the water in Sungai Johor, the Straits of Johor and the areas around Pulau Tekong. Let me just explain the colours shown on this image. The red colour depicts the land mass. The white colour depicts puffs of clouds. The blue water denotes the cleaner waters and the cloudy water denotes silty or muddy waters.

On the slide provided to the Court, was a close-up shot of the estuarine environment near Pulau Tekong taken in October 1998, before reclamation. Ms. Cheong asked the court to *"Notice the silty waters discharging from Sungai Johor and its tributaries southwards towards Pulau Tekong and the Straits of Johor. The area in white shows land clearance activities such as those at Tanjung Langsat in Malaysia"*.

Other images were also presented to the court: "This next slide is another close-up shot of land clearance activities at the eastern end of Tanjung Langsat in Malaysia. We see silty waters being discharged into Sungai Johor. Land clearance activities in Malaysia appear to contribute to discharge of silt in the Straits of Johor. In addition, Malaysia's own report by Delft Hydraulics called The Hydraulic and Environmental Impact Assessment for the Straits of Johor—highlighted that untreated waste water is being discharged by Malaysia into the rivers and the Straits of Johor. This slide shows a perturbed environment in the waters in Sungai Johor and around Pulau Tekong in April 2000, before reclamation took place. Hence, it can be seen that the waters around Pulau Tekong and the rivers in Malaysia are generally turbid all year round. This was the situation even before reclamation works at Pulau Tekong started in early 2001".

In its judgement in 2003 ITLOS directed Malaysia and Singapore to cooperate and enter in consultation, to establish a group of independent experts with the mandate to find solutions to solve the dispute. Also, ITLOS directed Singapore to prevent irreparable prejudice to the rights of Malaysia during its land reclamation.[120]

---

[118] ITLOS/Press 80, Press Release, 5 September 2003, https://www.itlos.org/fileadmin/itlos/documents/press_releases_english/PR_No.80.pdf, (accessed 27.09.2019).

[119] International Tribunal for the Law of the Sea, Minutes of Public Sittings, September–October 2003, Case concerning Land Reclamation by Singapore in and around the straits of Johor (Malaysia v. Singapore), Provisional Measures, https://www.itlos.org/fileadmin/itlos/documents/cases/case_no_12/Minutes_Case12.pdf, (accessed 24.09.2019).

[120] ITLOS, Order of 8 October 2003, List of Cases: No. 12, https://www.itlos.org/fileadmin/itlos/documents/cases/case_no_12/12_order_081003_en.pdf, (accessed 27.09.2019).

See also: ITLOS/Press 84, Press Release, Order in the Case Concerning Land Reclamation by Singapore in and around the straits of Johor (Malaysia v. Singapore), 8 October 2003, https://www.itlos.org/fileadmin/itlos/documents/press_releases_english/PR_No.84.pdf, (accessed 27.09.2019).

3.1.4 *Space Contributions to Provide Critical Infrastructure for Developing Countries*

It is important for the developing countries to be aware of and invest in space technology in order to take advantage of its benefits, including the promotion and support of sustainable solutions to achieve the SDGs and other socio-economic development. Space technology is now cheaper, more sustainable and commercially available, offers a wide area of application including agriculture, resources management, disaster management, communication, urban planning and security. In addition, there are plenty opportunities for international cooperation and also goodwill from developed countries to train personnel from the developing countries. The following examples will underline the related applications of space technology.

3.1.4.1 Small Satellite Programs

Small satellites are a fast-growing market and their production and launch will revolutionize the space industry. Small satellites have a bright future, having the possibility to develop areas of service in spatial and temporal coverage of the Earth.[121] Having been government and military dominated for many years, it is now reasonable to affirm that space access will be defined by an advanced level of commercialization. As costs come down, demand for launching small satellites and for benefitting from constellations of small satellites will rise. Technology development and the lower costs of equipment will open the market further for the commercialization of space. Satellites will become smaller, more flexible and reprogrammable, easier to replace, while the frequency of launches will bring costs down and be an incentive for innovation. Small satellites in LEO differ from GSO satellite by design, lifetime expectancy and lower cost of production. Mass production will make the work of designing a satellite redundant for a whole batch of satellites in a generation and will be replaced by innovative solutions.

Small satellites could provide large benefits to developing countries. They are relatively cheap to build, cheaper to launch and due to the advancement in technology they enable a variety of missions, from telecommunications to remote sensing to space science. Small satellites have a large area of activity. They could be used for remote sensing including EO, for global connectivity such as the Internet, for defence or even for deep space exploration. For example, small satellites of around 500 kg or less have been used by space agencies in earth science measurements in six interdisciplinary science focus areas: (i) atmospheric composition; (ii) carbon cycle and ecosystems; (iii) climate variability and change; (iv) weather; (v) water and energy cycle; (vi) Earth surface and interior.[122] The research of Neeck

[121] NLR, The bright future of small satellites, https://www.nlr.org/article/the-bright-future-of-small-satellites/, (accessed 21.08.2019).

[122] Neeck, S.P., Hammer, T.F. (auth.) Sandau, R., Roser, H.P., Valenzuela, A. (eds.), Small Satellites for Earth Observation, pp. 4, Springer Netherlands, 2008.

and Hammer included the analysis of some of the NASA missions involving small satellites, namely: the Ocean Surface Topography Mission (OSTM) will measure sea surface height; the Orbiting Carbon Observatory (OCO)—will provide space-based observations of atmospheric carbon dioxide (CO_2); the Glory and Global Precipitation Measurement (GPM) –will provide data to increase understanding of black carbon soot and other aerosols as causes of climate change and measure solar irradiance, while the Global Precipitation mission will provide measurements of global precipitation carrying active and passive microwave sensors.[123]

3.1.4.2 Current Use of Space-Based Telecommunication Infrastructure

This section will summarily analyse the telecommunication infrastructure of three developing countries, namely Mexico, China and Brazil. The purpose is to show the usefulness of satellite technology and the socio-economic benefits gained from investing in space capabilities.

Mexico

The Mexican Satellite System is known as MEXSAT. It comprises three GEO satellites, owned by the Mexican Government, MEXSAT-1, MEXSAT-2 and MEXSAT-3 and operated under the supervision of the Ministry of Communications and Transportation. MEXSAT-1 and MEXSAT-2 are used for mobile communication devices and operate in the electromagnetic frequencies of the L and Ku bands. MEXSAT-3 operates in the C and Ku bands. The main priorities of the Mexican Space Agency (Agencia Espacial Mexicana, AEM) programs are telecommunications for rural Internet, emergency response, remote fixed and mobile communications (MEXSAT) and EO for agriculture and defense purposes (SIAP). The AEM supports capacity development in small satellite technology, through technology development centres and via international cooperation.[124]

The MORELOS satellites launched since 1985 were operated between 1985 and 1998 by the Mexican commercial satellite operator, SATMEX.[125] Three planned MEXSAT satellites will coexist with SATMEX. The MORELOS-3 communications satellite was launched into orbit in October 2015 aboard a United Launch Alliance ATLAS V rocket. The MORELOS-3 is a Boeing built satellite, designed to operate in the GEO and be part of the Mexico's new MEXSAT system. The satellite is operated by the Telecomunicaciones de Mexico (Telecomm) on behalf of the Mexican

[123] Ibid. *supra* note Neeck and Hammer, Small Satellites for Earth Observation, pp. 5, Springer Netherlands, 2008.

[124] OECD (2018) The Space Economy in Figures How Space Contributes to the Global Economy. Country profiles: actors in the space economy, pp. 167.

[125] Matignon, L.G., Morelos1, The First Mexican Satellite, 2 June 2019, https://www.spacelegalissues.com/morelos-1-the-first-mexican-satellite/, (accessed 24.09.2019).

Government and is intended to provide L- band mobile communications to Mexico's Armed Forces and Federal Police. This satellite joins the existing MEXSAT constellation. MEXSAT was designed to support communication capabilities in Mexico, including in the rural and remote areas, while supporting Mexico's national security needs.[126] In December 2012, a smaller satellite called Bicentenario designed for the C- and Ku- bands connected Mexico's rural areas. Mexico has one more unoccupied slot for Geostationary Orbit that will remain available to Mexico until 2021.[127] "The Mexican government's decision to order three satellites from El Segundo, Calif.-based Boeing—one of those will be purchased by Boeing from Orbital Sciences Corp. of Dulles, Va. may be seen as going against a trend in Europe, the United States and parts of Asia. In these regions, governments are weighing whether and how to outsource their telecommunications services to the private sector, either by using commercial satellites or by placing government payloads aboard commercial satellites".[128] Mexico signed with Boeing a USD 1.032 billion contract, which includes two large L-band satellites and one smaller C- and Ku- band satellite from Orbital Sciences. In addition, Mexico acquired two Boeing provided control stations, network operations facilities and prototypes of user terminals.[129]

An analysis of the telecommunications sector in Mexico would further include the national context for telecommunications policies, general features of the regulatory regime, the telecommunications market and market participants, regulatory structures, regulatory institutions and processes, regulations and related policy instruments in the telecommunications sector, regulation of interconnection, regulation of prices, quality of service, resource issues, universal service obligations, international aspects, consumer protection, market performance (price, penetration rates, quality of service).[130]

Small satellites represent the key technology for space research and development in Mexico, mainly because of the lower cost of development and simple manufacturing. The possibility to use such technology results in the development of several projects for remote sensing, natural resource management, volcanic observations, climate monitoring and pollution measurement. Overall, small satellite projects have played a fundamental role in developing the space sector in Mexico, considered as a

---

[126]Howard, C.E., Boeing and industry partners launch communications satellite into orbit for Mexican communications, transportation, 2 October 2015, https://www.intelligent-aerospace.com/military/article/16538303/boeing-and-industry-partners-launch-communications-satellite-into-orbit-for-mexican-communications-transportation, (accessed 25.08.2019).

[127]Selding, P.B, Atlas 5 launches Mexico's Morelos-3 L-band Satellite, 2 October 2015, https://spacenews.com/atlas-5-launches-mexicos-morelos-3-l-band-satellite/, (accessed 26.08.2019).

[128]Selding, P.B., With $1B satellite Order, Mexican Gov't Looks to Change Satmex Relationship, 10 January 2011, https://spacenews.com/1b-satellite-order-mexican-govt-looks-change-satmex-relationship/, (accessed 24.09.2019).

[129]Selding, P.B., With $1B Satellite Order, Mexican Gov't Looks to change Satmex Relationship, 10 January 2011, https://spacenews.com/1b-satellite-order-mexican-govt-looks-change-satmex-relationship/, (accessed 23.08.2019).

[130]OECD, Mexico Regulatory Reform in the Telecommunications Industry 1998, http://www.oecd.org/regreform/sectors/2497378.pdf, (accessed 24.09.2019).

driver for the creation of the Mexican Agency.[131] Human Rights Watch has underlined that Mexico has registered little progress in improving human rights practices, with the security forces continuing to commit extrajudicial killings, enforced disappearances and torture. In particular, it is a known problem that migrants travelling through Mexico are frequently subject to abuses and human rights violations. There are many situations where these crimes remain unreported, one of the reasons being that the prosecutor's offices where reports can be made are far from the places where crimes are committed.[132] Small satellites could be a viable solution for Mexico to gather evidence for human rights violations.

China

China plans to develop a constellation of around 320 satellites for LEO, named HONGYAN. The satellites will be developed by the China Aerospace Science and Technology Corporation (CASC). The HONGYAN system translates as "wild goose" and the name likely comes from the fact that in Ancient China geese were used to deliver messages. The constellation of more than 300 satellites will provide global coverage and should be completed around 2025. This constellation will be operated by the Aerospace Dongfanghong Satellite Company, a CAST subsidiary.[133] The first HONGYAN prototype satellite launched for CASC's 300+ constellation went to a 1,100-kilometer orbit to test L- and Ka- band communications technologies.[134]

Another LEO constellation planned by China is HONGYUN meaning "rainbow cloud" previously dubbed XINGYUN. This constellation is being developed by CASIC, another state-owned company, sister to CASC and a major defense contractor that will build its own 156 satellite constellation in LEO.[135] The project was launched by CASIC in 2016 and aims to become operational around 2022. One experimental satellite was launched in orbit and four more satellites are expected to be launched up to the end of 2020. The constellation is an important part of

---

131 International Space University, White Paper Southern Hemisphere Space Studies Program, pp. 18-30, 2017, https://swfound.org/media/205747/isu-unisa_shssp_whitepaper_smallsats_8feb2017.pdf, (accessed 1.10.2019).

132 Human Rights Watch, Mexico Events of 2018, https://www.hrw.org/world-report/2019/country-chapters/mexico, (accessed 2.10.2019).

133 Jones, A., China to launch first Hongyan LEO communication constellation satellite soon, 2018, https://gbtimes.com/china-to-launch-first-hongyan-leo-communications-constellation-satellite-soon, (accessed 26.08.2019).

134 Gunter's Space Page, https://space.skyrocket.de/doc_sdat/hongyan-1.htm, (accessed 26.08.2019).

135 Jones, A., China to launch first satellite for HONGYAN global Internet satellite constellation on Saturday, 2018, https://gbtimes.com/china-to-launch-first-satellite-for-hongyan-global-Internet-satellite-constellation-on-saturday, (accessed 26.08.2019).

China's policy to commercialize space technologies and bring the Internet to about 600 million Chinese that are without access in remote and rural areas.[136]

Another project announced by China will consist of 72 satellites. The program will be implemented by a Beijing based private satellite company Commsat and will be funded by the Xi'an Institute of Optics and Precision Mechanics under CAS. Eight satellites are already being tested in orbit. The company aims to complete the global constellation by 2022. According to the available news, these satellites will be used for "wildlife protection, field emergency rescue, vehicle and ship monitoring and logistics tracing".[137] It seems clear that such satellites will have remote sensing and EO as their primary activity.

There are many satellites available for communications registered by China, around 17 satellite operator consortia. In addition to LEO mentioned above, the Asia Broadcast Satellite (ABS), the APT Telecommunications, AsiaSat and CAS must be mentioned.[138] These are mainly in GSO and are used for communication purposes, such as radio and television networks, backhaul, and direct broadcast.

Brazil

Mobile satellite systems are used in Brazil to provide communications over its extensive territory. Many communities live in remote areas and satellites are the major communication platform also in the Amazon Region of the North. Among the communication service provider companies in Brazil are Globalstar, Iridium and Inmarsat.[139] Embratel Star One also offers satellites for telecommunications.[140] Communication satellite operators include BRAMSAT, Embratel/Star One, INPE, Loral Skynet do Brazil and Telebras.[141]

Since 2003, Embratel has been owned by America Movil, a Mexican Telecommunication giant. Embratel Star One is a Brazilian communication satellite company and a subsidiary of Embratel. Active satellites are STAR ONE C and D, manufactured by Alcatel Alenia Space and Space Systems/Loral.

---

[136] Jing, M., China launches first satellite in bid to create global communications network and broaden net access, 2 January 2019, https://www.scmp.com/tech/policy/article/2180346/china-launches-first-satellite-bid-create-global-communications-network, (accessed 26.08.2019).

[137] ChinaDaily. China to launch constellation with 72 satellites for Internet of Things, 3 July 2019, http://www.chinadaily.com.cn/a/201907/03/WS5d1c543ea3105895c2e7b71d.html, (accessed 26.08.2019).

[138] Gunter's Space page, https://space.skyrocket.de/directories/sat_com_china.htm, (accessed 24.09.2019).

[139] The largest communication service provider companies in Brazil are Globalstar, Iridium and Inmarsat. The last two operate through distributors who sell their services and develop added value solutions. These companies offer voice services that include voice communication, broadband Internet, data services, tracking and monitoring.

[140] Tech in Brazil, Satellite Communication in Brazil, 29 June 2019, https://techinbrazil.com/satellite-communication-in-brazil, (accessed 24.09.2019).

[141] Gunter's space page, Spacecraft Communication Brazil, https://space.skyrocket.de/directories/sat_com_brazil.htm, (accessed 26.08.2019).

The market for small satellites is expected to grow in Brazil in the next few years. Brazil is seeking a partnership with U.S.A. to launch small satellites from the Alcantara base, near the Equator.[142] Small satellite policy in Brazil is being developed, focusing on benefits such as bringing innovation, new jobs and expanding the space industry. The Brazilian small satellite program includes missions for remote sensing, data collection, scientific, geoscience and technology and academia.[143] Human Rights Watch has raised the problem of migrants, refugees and asylum seekers in Brazil. Many Venezuelans have crossed the border into Brazil where they requested asylum. Also, another subject is the environment and land-related conflicts, where many rural residents fear reprisals from large landowners.[144] There are many opportunities to use the satellite imagery collected from small satellites in Brazil. For example, a Finnish company named ICEYE, which is building a constellation of small satellites of 100 kg, underlined that despite the often-cloudy skies over Brazil, the ICEYE satellite could monitor the progress of mud flows caused by a dam breach, with small satellites that are delivering radar image resolution under 1 m.[145]

3.1.4.3 Growing Demand in African Countries

The ITU estimated that at the end of 2018, 51.2% of the global population or 3.9 billion people were using the Internet. Further, from the ITU World Telecommunication/Information and Communication Technologies (ICT) indicators database, it can be stated that of all the ITU regions, Africa reported the strongest growth in the percentage of people using the Internet, which increased from 2.1% in 2005 to 24.4% in 2018.[146] According to the ITU, Africa is recording a shift to mobile-broadband services, especially among young consumers, as smartphones become more affordable. However, if comparing the growth in data services with the sharp decline in voice revenues, the progress is slow, and more investment is needed in network infrastructure to result in profitability. In general, investment is regarded as an essential factor for growth in telecommunications, necessary to create a sustainable market environment.

---

142 Boadle, A., Brazil space station open for small satellite business, 15 September 2018, https://www.reuters.com/article/us-space-brazil-usa/brazil-space-station-open-for-small-satellite-business-idUSKCN1LV007, (accessed 2.10.2019).

143 Rosa, A.C.G., Carvalho, H., Small Satellites: Challenges of the Brazilian National Space Law and Policy, September 2018, http://www.unoosa.org/documents/pdf/psa/activities/2018/Symposium_Brazil_BSTI/presentations/oral/S31_03_Ana_Cristina_Galhego_Rosa.pdf, (accessed 2.10.2019).

144 Human Rights Watch, Brazil Events of 2018, https://www.hrw.org/world-report/2019/country-chapters/brazil, (accessed 2.10.2019).

145 Berger, E., Small satellites begin to offer unrivaled detail in radar images of Earth, 8 September 2019, https://arstechnica.com/science/2019/08/small-satellites-begin-to-offer-unrivaled-detail-in-radar-images-of-earth/, (accessed 2.10.2019).

146 ITU News, New ITU statistics show more than half the world is now using the Internet, 6 December 2018, https://news.itu.int/itu-statistics-leaving-no-one-offline/, (accessed 25.09.2019).

For Africa, in strict relation to the expansion of the network infrastructure, the need for policy makers to adapt to this development and foster investment, without which it would be impossible to connect consumers and business to smart societies, was underlined.[147] To bridge the space divide and allow Africa to join the Internet-enabled economy, industry should adapt its business to the economic realities, support investment and provide service solutions that would strike a balance between costs and benefits of the service provider and consumer.[148] In particular, space technology can support the accessibility of the Internet and help develop the smart city concept, for example through large constellations of satellites to be deployed in LEO.[149]

The disparity in incomes between Africa and the other regions is noteworthy, which may explain the narrow range for telecommunication prices charged to consumers, somewhere between USD 5 (first quartile) to USD 13 (third quartile) per month.[150] The largest reductions in price of the mobile-cellular basket were recorded in Africa. In fact some African countries, such as South Sudan and Kenya, stand out for having the lowest prices worldwide for the mobile-cellular basket.[151] By contrast, the data volume price per GB in Africa was twice as high as in Europe and the CIS. In the case of Africa, the price per GB is the highest of all regions and therefore the report underlined that the challenge remains to extend affordable offers and plans including larger data allowances. For an overall picture over the region in relation to costs, it is worth mentioning examples, such as Rwanda, where packages with shorter validity periods (including time of use and targeting a specific user group) allow lower-income users to access the Internet at an affordable cost. Overall, statistics in Africa show that cost is not the main barrier to using the Internet, but it is rather a strong determinant of how much data is consumed. In relation to the price of the handset-based mobile-broadband basket, the two poles are as follows: the most affordable is in Mauritius, Seychelles, Gabon, Cabo Verde, South Africa and Nigeria, while it remains unaffordable to large segments of the population on the Central African Republic, DRC and Zimbabwe.[152]

Similar to the advent of the ICT, Africa is developing its capabilities in space and is becoming active in the satellite market segment. Ten African countries may

---

[147] ITU, Measuring the Information Society Report 2018, Vol. 1, pp. 91, https://www.itu.int/en/ITU-D/Statistics/Documents/publications/misr2018/MISR-2018-Vol-1-E.pdf, (accessed 25.09.2019).

[148] Kotze, C., Broadband Access for Remote Un-serviced Communities, pp. 1-16, in: Kotze, C., A Broadband Apparatus for Underserviced Remote Communities, Connecting the Unconnected, Springer, 2019.

[149] Hugbo, T., The Importance of Internet Accessibility and Smart City in Sub-Sahara African Region Through Space Technology, in: Froehlich, A., Embedding Space in African Society, The United Nations Sustainable Development Goals 2030 Supported by Space Applications, Springer 2019, pp. 105 – 113.

[150] Ibid. *supra* note ITU, Measuring the Information Society Report 2018, Vol. 1, pp. 108, 2018.

[151] Ibid. *supra* note ITU, Measuring the Information Society Report 2018, Vol. 1, pp. 109, 2018.

[152] Ibid. *supra* note ITU, Measuring the Information Society Report 2018, Vol. 1, pp. 123-124, 2018.

be highlighted for investing in using and launching satellites based on scientific, technological and military ambitions.[153]

Rwanda for example, has partnered with OneWeb, a UK based satellite operator that aims to operate a large constellation of satellite in LEO to provide the Internet in remote and rural areas. With this partnership, Rwanda hopes to connect rural and remote areas of the country to the Internet.[154] On the basis of this partnership, one of the first six satellites launched in February 2019 by OneWeb was named "ICYEREKEZO" by a Rwandan School in Nkombo Island in Lake Kivu.

The second satellite launched in space by Rwanda was a 3U CubeSat satellite named RWASAT 1 and was delivered to the International Space Station (ISS) by a Japanese cargo vehicle. The CubeSat was built under the auspices of the Rwanda Utilities Regulatory Authority (RURA) with the aim to develop a space engineering capacity in Rwanda.[155] The CubeSat carries a communication payload to collect and forward data from remote monitoring stations on the ground and also carries two cameras for Earth Observation.[156] The CubeSat may be regarded as a technology demonstrator, planned to be deployed from the ISS in the October to November 2019 time frame.[157]

Another example is Morocco, which launched its second EO satellite in November 2018. An Arianespace VEGA rocket launched the MOHAMMED 6-B satellite and placed it in the same orbit with its twin satellite MOHAMMED 6-A launched in 2017. Morocco now has a two-satellite constellation for EO.[158] The MOHAMMED 6-A satellite was developed by a consortium formed by THALES ALENIA SPACE and AIRBUS DEFENCE and SPACE. Since 2011, Morocco launched three satellites. Morocco has ratified the OST, the Rescue Agreement, the Liability Convention and the Registration Convention. The Royal Centre for Remote Sensing (CRTS) was established in 1989 and is responsible for most space activities in the country, addressing projects in fields such as: agriculture, forest area, town and spatial planning, oceanography, water resources, desertification, natural risks and disasters,

---

153 Pedroncelli, P., 10 African Countries Which Have Launched Satellites Into Space, 20 June 2019, https://moguldom.com/207725/10-african-countries-which-have-launched-satellites-into-space/, (accessed 25.09.2019).

154 Taylor, M.E., Rwanda launched first-ever satellite to connect remote schools to the Internet, 28 February 2019, https://face2faceafrica.com/article/rwanda-launches-first-ever-satellite-to-connect-remote-schools-to-the-Internet, (accessed 25.09.2019).

155 Gunter's Space Page, RWASAT1, https://space.skyrocket.de/doc_sdat/rwasat-1.htm, (accessed 25.09.2019).

156 Graham, W., HTV-8 launches on H-IIB to the ISS, 24 September 2019, https://www.nasaspaceflight.com/2019/09/jaxa-launch-htv-8-cargo-iss/, (accessed 25.09.2019).

157 Ibeh, J., Launch of Rwanda's 1st Satellite and Egypt's 8th Satellite Rescheduled for September 2019, 29 August 2019, https://africanews.space/rwanda-and-egypt-set-to-launch-satellites-into-space/, (accessed 25.09.2019).

158 Henry, C., Arianespace Vega launches second Moroccan Earth-observation satellite, 20 November 2018, https://spacenews.com/arianespace-vega-launches-second-moroccan-earth-observation-satellite/, (accessed 25.09.2019).

geographic information system, geospatial data and information, expertise and technical assistance and training activities.[159] Morocco is also very active and supports international cooperation, among its projects it signed a MoU with India in 2018, participates in the Space Climate Observatory (SCO), keeps strong relations with CNES, and participates in ExoMars program of ESA.[160]

Algeria could be reasonably described as one of the most active African countries in space technology, already having six satellites in orbit. Algeria launched its first satellite in 2002. ALSAT-1 was the first microsatellite launched under the Disaster Monitoring Constellation (DMC) and provides multispectral images for monitoring natural disasters. ALSAT-1B was launched in 2006 and was designed for agricultural and disaster monitoring.[161] In 2010, the Algerian remote sensing satellite ALSAT-2A was successfully launched into orbit, becoming the second satellite for Earth Observation. ALSAT-2B is similar. ALSAT-1 N was also launched in 2016 and contains amateur radio payloads. Alcomsat-1 is the first Algerian communication satellite. The telecommunication satellite was built in China and was launched by a Chinese rocket in December 2017.[162] The Algerian Space Agency (ASAL) was established in January 2002 and is in charge of the Algerian space program.[163] Algeria has ratified the OST, the Liability Convention and the Registration Convention.

There are many examples of the involvement of Algerian engineers and students in the building of the launched satellites. For example, AlSat-1 and AlSat-2 were built in partnership with AIRBUS Defense and Space and included the training of 30 space engineers, doctoral and master students. Also, Algerian students were involved in the project regarding the building of three CUBESATs in collaboration with the Surrey Space Centre from the United Kingdom.[164] Also, Algeria is an active member in international space programs, including the DMC and supports the implementation of the African Space Policy.

Ethiopia is one of the youngest space actors in Africa. The Ethiopian Space Science and Technology Institute (ESSTI) was established on 14 October 2016 with the

---

[159] Siebrits, A., Morocco, in: Froehlich, A., (Eds.), Integrated Space for African Society, Legal and Policy Implementation of Space in African Countries, Springer, 2019, pp. 171-201.

[160] Ibid. *supra* note Siebrits, A., Morocco, in: Froehlich, A., (Eds.), Integrated Space for African Society, Legal and Policy Implementation of Space in African Countries, Springer, 2019, pp. 171-201.

[161] Kameche, M., Benzeniar, H., Benbouzid, A.B., Amri, R., Bouanani, N., Disaster Monitoring Constellation using Nanosatellites, https://www.spacemic.net/abstracts/[Non-Finalist]%20Disaster%20Monitoring%20Constellation%20using%20Nanosatellites.pdf, (accessed 25.09.2019).

[162] Henry, C., Chinese Long March 3B launches Algeria's first telecom satellite, 10 December 2017, https://spacenews.com/chinese-long-march-3b-launches-algerias-first-telecom-satellite/, (accessed 25.09.2019).

[163] Adde, Y.A., Socioeconomic Benefits of Space Technology for Africa, IEEE-SEM, Volume 10, Issue 6, pp. 259, June 2019, http://www.ieeesem.com/researchpaper/Socioeconomic_Benefits_of_Space_Technology_for_Africa.pdf, (accessed 25.09.2019).

[164] Siebrits, A., Algeria, in: Froehlich, A., (Eds.), Integrated Space for African Society, Legal and Policy Implementation of Space in African Countries, Springer, 2019, pp. 113-143.

aim to enable the country to enter into space activities.[165] As a result of government-to-government collaboration between China and Ethiopia, Ethiopia is preparing to launch two satellites.[166] According to the public information available, the first satellite will be an EO satellite called ETRSS-1 that is expected to be launched from China in late 2019. The other satellite is a telecommunication satellite financed by China with USD 6 billion out of the USD 8 billion cost. Ethiopia also hosts the East Africa Regional Office of Astronomy for Development (EAf-ROAD), opened since 2014 under agreement with the International Astronomical Union (IAU).[167] Ethiopia has signed the OST and ratified other two agreements regarding international telecommunications. As part of its efforts to have a space program, a privately funded astronomical observatory was built by the Ethiopian Space Science Society (ESSS), which enabled the nation also to involve the universities in the space activities.[168] Overall, it was mentioned that scarce funding, political visibility and weak industrial sector are key challenges to be overcome in order to further develop the space program.

Kenya launched its own first satellite in May 2018. The CubeSat named 1KUNS-PF was built at the University of Nairobi in partnership with the Japanese Space Agency (JAXA) and was launched from the Japanese module of the ISS. The CubeSat was designed for weather forecasting, food security mapping, livestock and wildlife monitoring and disaster management.[169] It is noteworthy to also mention that Kenyan space history includes a launch site, launching nine satellites and twenty research rockets between 1967 and 1988. The first satellite launched from Kenya was NASA sponsored.[170] Kenya has ratified the OST and the Liability Convention. Also, the University of Nairobi has a department of geospatial and space technology, offering programs on astrophysics and remote sensing. Overall, Kenyan space activities depend on collaboration with other countries and international organizations.[171]

Egypt launched its first satellite in 2007. However, Egypt has big aspirations in space activities and has already put several satellites in orbit, among which is the NILESAT satellite used for communication and launched between 1998 and 2010. In February 2019 Egypt launched another satellite for EO purposes, the EGYPTSAT-A

---

165 ESSTI Ethiopian Space Science & Technology Institute, About us – Establishment of EESTI, http://etssti.org/home-2/, (accessed 25.09.2019).

166 Ibid. *supra* note Adde, Y.A., Socioeconomic Benefits of Space Technology for Africa, IEEE-SEM, Volume 10, Issue 6, pp. 260, June 2019.

167 ESSTI Ethiopian Space Science & Technology Institute, About EAf-ROAD, http://etssti.org/ea-road/about-eaf-road/, (accessed 25.09.2019).

168 Ogunyinka, A.O., Ethiopia, in: Froehlich, A., (Eds.), Integrated Space for African Society, Legal and Policy Implementation of Space in African Countries, Springer, 2019, pp. 255-265.

169 Dahir, A.L., Kenya heads into space with the launch of its first home-designed cube satellite, 11 May 2018, https://qz.com/africa/1275698/kenya-to-launch-first-satellite-into-space/, (accessed 25.09.2019).

170 Ibid. *supra* note Adde, Y.A., Socioeconomic Benefits of Space Technology for Africa, IEEE-SEM, Volume 10, Issue 6, pp. 259, June 2019.

171 Martens, B., Kenya, in: Froehlich, A., (Eds.), Integrated Space for African Society, Legal and Policy Implementation of Space in African Countries, Springer, 2019, pp. 157-171.

high-resolution, which is the third EO satellite, cooperated and developed with Russia. The remote sensing satellites delivered to Egypt were mainly used in monitoring natural disasters, geological resources, agricultural productivity, archaeological purposes and aquatic life monitoring.[172] It was announced in August 2019 that two CUBESAT were built entirely by a team of engineers working for the National Authority for Remote Sensing and Space Sciences (NARSS) Egypt and funded by the Academy of Scientific Research and Technology. The NARSS built CubeSat (NARESCOPE-1) was launched to the ISS in August 2019.[173] The NARESCOPE-2 was launched in September 2019 containing new imaging technology, with high-storage capabilities, tested at the Kitakyushu University in Japan. The third CubeSat should be completed in six months' time and is planned for launch during the third quarter of 2020.[174]

Ghana's space program was established in 2011 and was operated by the Ghana Space Science and Technology Center (GSSTC). In 2016, the Ghana Space Agency (GhSA) was created. Ghana launched its first satellite in 2017. In particular, GHANASAT-1 was a CubeSat developed by students at All Nations University in Koforidua and was deployed from the ISS. GHSANASAT-1 took two years to build and the project was supported by JAXA. The satellite was developed for mapping purposes and aimed to build capacity in space science and technology.[175] Ghana has signed the OST, the Rescue Agreement and Liability Convention. As part of its international programs, Ghana participates in the South African Largest Telescope project (SALT) and the Square Kilometre Array (SKA). Overall, the GhSA could support solutions to the challenges Ghana is faced such as illegal gold mining, climate change, natural disasters and economic development.[176]

Angola launched its first satellite in December 2017, but the ANGOSAT-1 failed in orbit four months after the launch. ANGOSAT-2 will be the replacement for the failed satellite and will be financed partly by the insurance and partly by Russia.[177] The new telecommunication satellite is planned for launch in 2020 and will include 16 transponders in C-band and 6 in Ku-band. Without any of the space treaties to be signed or ratified, Angola approved a National Space Policy in 2017, promoting the development of spatial infrastructure, training of specialists and exploration

[172] Ibid. *supra* note Adde, Y.A., Socioeconomic Benefits of Space Technology for Africa, IEEE-SEM, Volume 10, Issue 6, pp. 260, June 2019.

[173] Al-Youm, A.M., Egypt construct its first-ever two satellites, 02 August 2019, https://egyptindependent.com/egypt-constructs-its-first-ever-two-satellites/, (accessed 25.09.2019).

[174] Sakr, T., Egypt launches its 2nd satellite Narescope-1 from Japan, 19 September 2019, https://ww.dailynewssegypt.com/2019/09/19/egypt-launches-its-2nd-satellite-narescope-1-from-japan/, (accessed 25.09.2019).

[175] BBC, Ghana launches its first satellite into space, 07 July 2017, https://www.bbc.com/news/world-africa-40538471, (accessed 25.09.2019).

[176] Gasela, M., Ghana, in: Froehlich, A., (Eds.), Integrated Space for African Society, Legal and Policy Implementation of Space in African Countries, Springer, 2019, pp. 231-243.

[177] Adamowski, J., Angola eyes new satellite as African space race accelerates, 12 June 2018, https://spacenews.com/angola-eyes-new-satellite-as-african-space-race-accelerates/, (accessed 25.09.2019).

technologies. Overall, Angola is investing in space infrastructure for the inclusion of ANGOSAT-2 and hopes to develop telecom business opportunities once the satellite is successfully deployed in orbit.[178]

South Africa launched its first satellite for EO in 1999, called SUNSAT, which is the first ever launched African satellite.[179] The South African National Space Agency (SANSA) was then established in 2010. This is the reason why South Africa could be considered the pioneer of harnessing space technology in Africa. South Africa is also known for its advancements in the field of astronomy, with SALT near Sutherland being the largest optical telescope in the southern hemisphere. In December 2018, South Africa launched the ZACube-2, which was described as Africa's most advanced CubeSat. The CubeSat was designed to provide remote sensing and communication services in the region, with a focus on natural disasters such as monitoring wildfires.[180] South Africa has ratified two international outer space treaties, namely the OST and the Liability Convention. The SANSA Space Operations Directorate provides a wide range of services for EO satellites, including raw image applications, processing services and fire scan mapping, while the SANSA EO Services is acting as a data reseller.[181] Consequently, space-related activities and space-derived products play a significant role in South Africa's economy.[182]

Finally, the Nigerian program has to be highlighted for its advancements. Nigeria is the largest economy in Africa and along with South Africa, the Nigerian space program can be considered the most advanced in Africa.[183] The National Space Research and Development Agency (NASRDA) was established in 2001. Nigeria launched its first EO satellite NIGERIASAT-1 in 2003. Since then, Nigeria has launched four more satellites, with only one failure. In 2007, Nigeria launched the first African communication satellite, NIGCOMSAT-1 built and launched from China, which

---

[178] Woodgate, S., Angola, in: Froehlich, A., (Eds.), Integrated Space for African Society, Legal and Policy Implementation of Space in African Countries, Springer, 2019, pp. 243-255.

[179] Ngcofe, L., Gottschalk, K., The growth of space science in African countries for Earth Observation in the 21st century, S Afr J Sci. 2013;109(1/2), Art. #a001, 5 pages. http://dx.doi. org/https://doi.org/10.1590/sajs.2013/a001, 2013.

[180] Isilow, H., S. Africa launches its most advanced nanosatellite. Described as Africa's most advanced cube satellite, ZAcube-2 is South Africa's 2nd nanosatellite to be launched, 27 December 2018, https://www.aa.com.tr/en/africa/safrica-launches-its-most-advanced-nanosatellite-/1350406, (accessed 25.09.2019).

[181] Alberts, A., South Africa, in: Froehlich, A., (Eds.), Integrated Space for African Society, Legal and Policy Implementation of Space in African Countries, Springer, 2019, pp. 59-81.

[182] Ibid. *supra* note Alberts, A., South Africa, in Froehlich, A., (Eds.), Integrated Space for African Society, Legal and Policy Implementation of Space in African Countries, Springer, 2019, pp. 59-81.

[183] Giles, C., Africa leaps forward into space technology, 16 May 2018, https://edition.cnn.com/2017/08/10/africa/africa-space-race/index.html, (accessed 25.09.2019).

was replaced in 2011 with NIGCOMSAT-1R.[184] In 2018, Nigeria agreed on a USD 550 million satellite deal with China, which will allow China to take an equity stake in NIGCOMSAT.[185] Nigeria has ratified four UN treaties on outer space, namely the OST, the Rescue Agreement, the Liability Convention and the Registration Convention. Also, Nigeria has begun the process of building a National Geospatial Data Infrastructure (NGDI), using satellite data from the three Nigerian satellites currently in orbit in order to meet the Sustainable Development Goals. Overall, Nigeria contributes towards the implementation of the African Space Policy and Strategy and places a strong emphasis on three key areas namely EO, satellite communication and space science and astronomy.[186]

In 2018, the African Space Agency (ASA) was constituted as an organ of the African Union (AU) and dedicated to "promoting, advising and coordinating the development and utilization of space science and technology in Africa (…)".[187] In 2019, Egypt won the bid to host the Headquarters of ASA.[188] The creation of ASA had been long discussed but in 2013 ASA was still considered premature. Peter Martinez backed by Carla Sharpe from the Square Kilometre Array (SKA)[189] cautioned that countries that have not yet participated in space endeavours, need first to develop their own capabilities to participate fully in the development of ASA.[190] It could be reasonably highlighted that this forewarning was fulfilled recently. With more than ten countries developing national space capabilities and cooperating internationally to launch satellites, space activities are becoming more visible in Africa.

A pan-African space program can be regarded as necessary in respect of the advancements and historic investment in space activities in Africa. The investment in space technologies and activities in Africa, especially the development of satellite EO and telecommunications programs contributes to the achievement of the UN SDGs and would increase the ability to plan and make strategic decisions for leaders

---

184 Nkordeh, N., Bob-Manuel, I., Oni, O., A Detailed Over-View of Satellites Launched by National Research and Development Agency from 2001-2016, Proceedings of the World Congress on Engineering and Computer Science 2017 Vol 1 WCECS 2017, San Francisco, USA, October 25-27, 2017, http://www.iaeng.org/publication/WCECS2017/WCECS2017_pp37-41.pdf, (accessed 25.09.2019).

185 Reuters, Nigeria agrees $550 million satellite deal with China, 3 January 2018, https://www.reuters.com/article/us-nigeria-satellite-china/nigeria-agrees-550-million-satellite-deal-with-china-idUSKBN1ES1G0, (accessed 25.09.2019).

186 Eriksen, C., Nigeria, in: Froehlich, A., (Eds.), Integrated Space for African Society, Legal and Policy Implementation of Space in African Countries, Springer, 2019, pp. 97–113.

187 Article 2 of the Statute of the African Space Agency, adopted 29 January 2018, https://au.int/sites/default/files/treaties/36198-treaty-statute_african_space_agency_e.pdf, (accessed 25.09.2019).

188 Writers, S., Egypt to Host African Space Agency's Headquarters—Foreign Ministry, 11 February 2019, http://www.spacedaily.com/reports/Egypt_to_Host_African_Space_Agencys_Headquarters___Foreign_Ministry_999.html, (accessed 25.09.2019).

189 Giles, C., Africa leaps forward into space technology, 16 May 2018, https://www.cnn.com/2017/08/10/africa/africa-space-race/index.html, (accessed 27.09.2019).

190 Ibid. *supra* note Ngcofe, L., Gottschalk, K., The growth of space science in African countries for Earth Observation in the 21st century, S Afr J Sci. 2013;109(1/2), Art. #a001, 5 pages. http://dx.doi. org/https://doi.org/10.1590/sajs.2013/a001, 2013.

in Africa.[191] In addition, an in depth analysis of Africa should consider its political organization, as many political challenges influence progress, including space. In respect of any activity, including space activities and cooperation across the continent, the African Union is the most important political organization and can be subdivided into eight intergovernmental organizations.[192]

As a concluding thought in regard to the advancement of space capabilities in Africa and in the AU, it could be highlighted that while the trend in Africa is to intensify space programs and develop its own capabilities of building satellites, such efforts are mostly in regard to cheap CubeSats. The bigger and more expensive satellites such as GEO satellites are developed and launched with the help of foreign investment, from nations including China, Russia, Japan and the EU. For this reason, Africa should also invest in human capacity development and work jointly with developed countries to benefit from the reward of investing in space capabilities.

191 De Waal Alberts, A., The Possible Beneficial Effect of Using Small Satellite Technology to Promote the Achievement of the UN Sustainable Development Goal of Poverty Reduction Specifically on the African Continent, in: Froehlich, A., Embedding Space in African Society, The United Nations Sustainable Development Goals 2030 Supported by Space Applications, Springer 2019, pp. 181–195.

192 Siebrits, A., Africa and the Space Arena, in: Froehlich, A. (Eds.), Space Supporting Africa, Volume 1: A Primary Needs Approach and Africa's Emerging Space Middle Powers, Springer, 2019.

Conclusions on Human Rights and Outer Space

The first chapter of this research discussed human rights in general. The protection of human rights in international law dates from the end of the Second World War and the creation of the UN. As was shown, the UDHR and the UN Charter are the basic texts for the protection of human rights. In addition, the OHCHR is the leading UN entity on human rights. With the emerging technologies, the future of human rights is not easy to predict, but nevertheless, human rights in international relations have to be protected. The positive contributions of technology should support human rights and social benefits, and any use of space technology that could potentially interfere with the classic meaning of privacy should justify the actions, such as proving criminal offences or other illegal actions. The first chapter outlined why the SDGs are particularly relevant for human rights. In terms of peace, justice and human rights, SDG16 addresses the promotion of peaceful and inclusive societies for sustainable development. SDG17 promotes the strengthening of the various means of implementation and revitalization of the global partnership for sustainable development.

In addition, it was concluded that data from EO satellites can help in gathering evidence of international crimes, especially in areas where access may otherwise be difficult, if not impossible. Topical discussion on human rights in relation to NewSpace activities is related to satellite-based data such as images and video from remote sensing. With the advent of space commercialization and low costs both for launching and manufacturing but also for services and finite products, the technology will become increasingly accessible to private companies or even individuals. The intrusion could be dangerous and is expected to become more advanced with the advent of artificial intelligence processing big data from space. In relation to privacy, the regulatory work is characterized by complexity and by different approaches to defining it and determining its focus. The fundamental rights of each individual require protection in every scenario of the future development of technology. Space could contribute to the protection of HR. Space addresses advanced, state-of-the-art technology that is constantly evolving. Directly proportional with the advancement of technology, the regulatory system should also take a leap forward and provide the necessary legislation to allow the use of satellite imagery to document HR abuses and humanitarian disasters.

A. Froehlich and C. M. Tăiatu, *Space in Support of Human Rights*,
Studies in Space Policy 23, https://doi.org/10.1007/978-3-030-35426-8

It was analyzed that human factors play a critical role in the design and interpretation of remotely sensed imagery. Digital image processing involves the manipulation and interpretation of digital images from remotely sensed data using digital image-processing software techniques. The challenges for the utilization of satellite remote sensing data include possible abuses in areas of privacy, human rights and public order. One of the most important legal aspects in relation to data collection is privacy. The challenges refer to what kind of data can actually be collected and distributed and who can have access to it. The use of satellite-collected images in human rights investigations reflects a number of opportunities but also challenges and risks. The advantages of satellite imagery include their capability to cover much greater areas, have the potential to offer real-time streaming and global coverage, with better prices. Because all information is digital, it can be easily transmitted. Satellites are outside the sovereign borders of each country which is why no authorization is needed from the sensed states, and satellites can be used to capture information from the most remote areas of the Earth. The first chapter concluded that satellite imagery is considered as a unique tool necessary in solving certain cases such as boundary disputes or environmental issues. The protection of human rights is an important objective to which satellite imagery could contribute in detecting human rights violations.

The second chapter of this research analyzed the usefulness of satellite imagery from a practical perspective. Cases were presented to indicate that satellite imagery plays an important role in updating and extracting land related information, and can provide valuable evidence such as a historical record of the areas that are subject to changes over time and prevent conflicts over boundaries which may involve violations of human rights. Advances in space-based remote sensing have revolutionized the field of cadastral surveying and mapping because satellite imagery can provide data for mapping applications and can be used as an alternative or a support to traditional land surveying. Satellite data can support the procedures for base mapping which is compulsory for the analysis of the documentary material for decision making and implementation.

A mechanism to protect human rights was discussed, including the work of non-governmental organizations for human rights. Various cases prove that satellite imagining is necessary to prove infringement of the law and violations of human rights. Mass displacement has been evidenced by satellite images. In general, the procedure to locate, identify and interview individuals is not easy in conflict zones or in other areas with high-risk for foreign scientists or reporters. To minimize the likelihood of terrorist attacks or other unlawful actions against foreign interviewers or the safety and security of the interviewee, to have instant access to information and real time images or access to previous images, satellite technology may be used to support the work in the field in certain areas and periods and document remote locations where the risk of conflict is high.

The combination of satellite images, geographic information/earth observation systems and global navigation satellite systems is necessary to support the work of local and international human rights organizations. The capacity of satellite images to improve the probatory procedures in the judicial system and prove violations of human rights and international humanitarian law was illustrated. However, there are various challenges in the interpretation of satellite images necessary for the prosecution of criminals.

Satellite images have proven useful in monitoring and protecting human rights in case of refugees and/or internally displaced persons. Interoperable data are necessary to support local action, particularly satellite-based data. It was also shown how intelligence agencies monitor arms flows including large-scale and high-resolution satellite or aerial images. International arms flows monitoring through satellite images could be included as part of the new innovative tools for assessing arms flows. Limitations include the significant resources needed for large-scale collection and processing of such data. More effective data collection efforts require careful consideration of the new tools.

The second chapter also analyzed how satellite imagery was crucial in the "Oil Platforms" case. It was also underlined that observations from satellites have become indispensable for forecasting the weather. Satellite imagery could also be a useful tool in waste management. It was underlined that space technology added value for tracking illegal dumping activities by using EO and GNSS satellite data. The effectiveness of satellites for oil spill surveillance has been proven many times. Real time deforestation detection through satellite images could be among the game changing solutions to save the Amazon Rainforest. It was further shown that the UN Principles on Remote Sensing distinguish between raw data and processed data. In this regard, satellite data as evidence in court faces questions about the authenticity of the processed data. The lack of consistency and standardization in processing data was discussed—it requires a competent authority with certification power. As per treaty compliance requirements, there is the need for treaties to specifically provide for use of satellite data as a tool to monitor compliance.

Chapter 3 referred to the use of satellite data in courts. This chapter presented cases from the ICJ and the ICC, concerning violations of human rights. The ICJ is the principal judicial organ of the UN. The jurisdiction of the ICJ comprises all cases which the parties refer to it and all matters specially provided for in the UN Charter or in treaties and conventions in force. The ICJ has contributed to the development of human rights by providing solutions to escalating international problems, condemning violations of human rights such as terrorism and the unilateral use of force. Satellite data is important evidence for the ICJ to solve disputes in accordance with international law. The third chapter presented examples of such ICJ cases including boundary disputes.

The ICC is a court of last resort for the prosecution of serious international crimes, including genocide, war crimes, and crimes against humanity. The third chapter presented examples of ICC cases and showed the importance of collaborating with a satellite imagery expert in interpreting satellite images provided by the parties. The ICC cases showed the need to provide explanations to judges and prosecutors regarding the procedure to produce and interpret satellite images used as evidence. Satellite imagery as evidence for human rights violations was discussed in a blog post as part of the International Justice Monitor's technology for truth series. Also, the ICC blog mentions the phenomena that can be detected with the help of satellite imagery including: "cratering consistent with bombardment; the destruction of civilian dwellings;

targeting of humanitarian and other forms of internationally protected infrastructure; the locations, size, and composition of displaced persons camps; excavations consistent with mass graves; and other probative phenomena". Satellite evidence could provide critical circumstantial evidence.

ITLOS was also presented in the third chapter, as an independent judicial body established by the UNCLOS. ITLOS cases were analyzed in this chapter in relation to satellite imagery underlining the benefits to developing countries.

Small satellites are cheap to build, cheaper to launch and advancements in technology could make them a viable solution to successfully replace larger and more expensive satellites, making them advantageous for developing states. Small satellites could be used for Remote sensing including EO, for global connectivity such as the Internet, for defence or even for deep space exploration. Chapter 3 presented the current use of the telecommunication infrastructure of three developing countries, concluding that small satellites could become part of their space infrastructure. Chapter 3 further focused on the growing demand in African countries, where small satellites technology could support the accessibility of the Internet and help develop the smart city concept. This chapter contained examples of space activities in African countries, highlighting the trend in Africa to intensify space programs.

Overall, the new trend for satellites is to deploy smaller, cheaper to build and easily replaceable satellites in LEO. The reasons for investing in deploying constellations of satellites include real time global coverage, low latency that will support innovation in telecommunication technologies on Earth, and also security reasons (military, national, international). The innovative technology used for low cost small satellites will allow them to be useful for a wide range of activities including commercial space, such as providing telecommunications and Internet, remote sensing for EO, video and satellite imagery in real-time, and military uses such as highly sensitive and capable radars able to support national defense systems and detect the new generation of missiles.

Space commercialization and democratization of space means the benefits of the outer space will become largely available to private entities and individuals who will operate their own satellite systems and provide services back to Earth. Once small satellites LEO constellations are able to provide global coverage near real-time, high resolution video and imagery, there will be the problem of respecting human rights.

The first question that is being raised currently is whether there is any specific piece of legislation that deals with protecting personal data from high resolution remote sensing. Clearly, improvements are necessary to specifically deal with satellite imagery from private entities. There are currently legislative solutions that could restrict the use of satellite imagery, but would this be sufficient for future development and expansion of private entities in space? This question should be further be analysed by policy makers and regulators to find solutions.

The second question that is raised to a greater extent is about using satellite imagery as evidence in courts, for example to prosecute those responsible for human rights violations, including in some of the conflict areas in Africa. As was presented in the research, satellite imagery is used widely and brings certain benefits such as wide coverage, costs, historic analysis and availability in conflict areas where there is restriction of movement. However, satellite imagery is gathered as raw data. For the final product to be produced and provided to courts, a series of technical analysis and interpretations are needed. There are currently no agreed standards for such technical support, and there is limited number of experts at UN level while courts could be reluctant to give such evidence its necessary value. Therefore, the standardization of satellite imagery analysis is urgently needed, and training must be provided to courts for them to better understand how useful such evidence is, how it is gathered from satellites, and how much availability there is to gather such information.

Certification of satellite data is necessary to prevent the final product from being intentionally distorted. Once a special authority is able to certify and authenticate the final product, such data could become widespread in investigations as a solid probatory means.

The manufacturer's authorised representative in the EU is Springer Nature Customer Service Centre GmbH, Europaplatz 3, 69115 Heidelberg, Germany. If you have any concerns regarding our products, please contact ProductSafety@springernature.com

Printed and bound by CPI Group (UK) Ltd, Croydon, CR0 4YY
30/09/2026
02225645-0014